Beyond Mechanization

Beyond Mechanization:

Work and Technology in a Postindustrial Age

Larry Hirschhorn

The MIT Press
Cambridge, Massachusetts
London, England

© 1984 by The Massachusetts Institute of Technology

This book was set in Baskerville by The MIT Press Computergraphics Department and printed and bound by Halliday Lithograph in the United States of America.

Library of Congress Cataloging in Publication Data

Hirschhorn, Larry.
    Beyond mechanization.

    Includes bibliographical references and index.
    1. Machinery in industry. 2. Man-machine systems.
3. Cybernetics. 4. Division of labor. 5. Work design.
I. Title.
HD6331.H5   1984   338'.06   84-947
ISBN 0-262-08142-3

To the memory of my parents
and to Mr. Leo Friedmann

Contents

Preface

Many people supported my writing efforts. Tom Gilmore and the staff of the Management and Behavioral Science Center at the University of Pennsylvania provided a collegial and stimulating setting for my research and consulting work. Hans Van Beinum and the staff of the Ontario Quality of Work Life Center helped me understand the complexities of sociotechnical settings. Fred Block examined the manuscript with a critical but sympathetic eye at key points in its evolution; his advice was invaluable, as was the editorial assistance of Lou-Ann Matossian. Marla Isaacs, my wife, showed me that writing is only one among many activities through which we develop. She provided the critical balance required as drafts found their circuitous way to final copy.

The Rockefeller Foundation for the Humanities supported my research with a one-year fellowship.

Beyond Mechanization

Robots can't run factories. The common notion that computers eliminate the need for human skill and judgment is wrong. We are still mesmerized by the nineteenth-century image of technology: mechanics and engineers together creating machines and machine systems that operate as if under immutable natural laws. Sigfried Giedion, the master historian of mechanization, evokes this image when describing an automobile frame factory: "The material is worked upon and moves back and forth through the factory on the most varied types of conveyor systems in an uninterrupted process. . . . Here, extremely precise time charts guide the automatic cooperation of instruments which, like the atom or a planetary system, consist of separate units, yet gravitate about one another in obedience to their inherent laws."[1]

Yet such machinery was extremely rigid. If the conditions of production (levels of heat, vibration, or intensity of use), the qualities of inputs, or the specifications for outputs changed, these machine systems were costly to adjust. Skilled and unskilled labor had to be used to keep the machinery adaptable.

Cybernetic technology seemed to fulfill the unkept promise of mechanical technology. Engineers could design control systems that flexibly responded to unpredictable changes in ambient conditions. Sensors could detect slight changes in vibrations, temperature, pressure, and speed and automatically adjust the machine's operation. The controlling computer could be reprogrammed so that the machine's cutting tools would vary their motions as the qualities of inputs or outputs changed. Workers had only to tend the machine.

But we are now learning that this cybernetic image of the perfect machine is as utopian as was its nineteenth-century counterpart. The new cybernetic machines create new sources of error and failure with which only skilled workers, ready to learn and adapt to new production conditions, can contend. The recent failures of nuclear reactors have been most revealing. Engineers have been unable to design complete backup and fail-safe systems. The complexities of the reactors and the unanticipated interactions between the control systems and the pumps and pipes have introduced new modes of failure. Worker mistakes during reactor breakdowns have demonstrated how poorly we understand the skills and job designs required for such settings.

In cybernetic settings workers must control the controls. To do so, they cannot merely become competent at a fixed and predictable set of tasks. Unlike nineteenth-century craftsmen, they cannot be enclosed in a limited view of the production process. Instead they must be able to survey and understand the entire production process so that they are ready to respond to the unpredictable mishap. Because the machine design is imperfect, they must continue to learn about its operation and contribute to its development. Machine systems can no longer be regarded as fixed and perfect but rather as shaped by contingencies. Workers learn and machine systems evolve. As products and processes change at an increasing rate and specialized markets replace mass markets, it becomes all the more necessary to understand this. Workers must not only control the controls when unpredictable breakdowns occur; they must increasingly monitor the controls as the software and hardware is redesigned to produce new products in new ways and with new inputs. In the mechanical age the semiskilled worker performed rote tasks; the skilled worker, varying tasks within a fixed framework. The postindustrial worker, in contrast, performs developmental tasks, operating at the boundary between old technical realities and emergent ones.

New factory settings have been devised in the last decade that reflect the postindustrial conception of work and technology. In a polypropylene chemical plant, the computer was not designed to control all the feedback loops of the production system. Chemical engineers could not fully represent the dynamics of the chemical process in mathematical terms, and previous attempts at computer control in other plants had resulted in significant underproduction. In the new plant the computer system was designed to facilitate worker learning. "The computer answers queries put to it by the operating personnel regarding the short-run effects of variables at various control levels, but decisions are made by the operators. Operating personnel are provided with technical calculations and economic data, conventionally only available to technical staff, that support learning and self-regulation. In this manner operator learning is enhanced."[2]

The learning design for computer control is matched by a pay-for-learning system of job classification. Workers get paid more as they learn more skills, and everyone can reach the top plant rate. The regular progression requires the "acquisition of skill in six process areas and two other skill areas to reach the top." The laboratory progression requires demonstrable skill in "three process areas and thirty-six lab modules to reach the top rate."[3] Workers train and test one another. This technical and social design is based on an appreciation of the new technologies: "The recognition that higher levels of technology, frequently accompanied by very large capital expenditures, increase the dependence of organizations on their workers, rather than the opposite as predicted by engineers, was crucial to the design process."[4]

Such settings must grow in numbers and significance. We must transcend the mechanical conception of work and technology if we are to utilize fully our new postindustrial technologies. Yet there is much resistance to change. The managers, union leaders, designers, engineers, and educators who together shape our new technologies operate under traditional industrial assumptions. Many still believe in dividing up the work, creating semiskilled jobs, and planning the work flow entirely from the back office. They argue that such designs reduce labor costs, minimize training time, and permit managers to substitute one worker for another. But these arguments lose force as capital costs rise in relationship to labor costs, learning ability becomes more important than past training, and a worker's tacit understanding of a particular machine system becomes more important than his general knowledge. Paradoxically, just as we are developing technologies characterized by the utmost mathematical abstraction, we must increasingly rely on informal learning and the ability to deal with the unpredictable.

We lack a full understanding of the new technology. To appreciate its potentials and limits, we must see how it emerged and how it is unfolding. We must examine

• the technical foundation of mechanical machine systems;

• the transition from mechanical machine systems to cybernetic ones, the differences between the two, and the current tensions between cybernetic machine systems and older methods of job design and industrial management; and

• experiments in designing work settings that meet the potentialities and requirements of cybernetic technologies and facilitate learning.

This examination will reveal three critical discontinuities:

• the discontinuity in machine design,

• the discontinuity in the relationship between workers and machines, and

• the discontinuity in the relationships among people at work.

Technology alone cannot determine work and organizational design, which are also shaped by social and political interests. But technology can set the limits within which design decisions are made. The new technology presents new opportunities and problems for work design. Moreover, like one drug that makes another more powerful, a technology can potentiate latent cultural trends. Control system failures may help to bring out in the culture a developmental concept of the self, a concept that leads people to seek out learning opportunities throughout their lives.

Yet the limits of technology as a determining force are the boundaries of this book as well. The book describes a potential system of technology and work, drawing on current experiences and conundrums; it does not estimate the likelihood that such a system will emerge. Social and political forces will shape a transition period in which old social contracts, organizational designs, and power relations may come into conflict with the potentialities and requirements of the new technology. The picture of the new technology painted here may help us to assess the present state of our work systems, but it cannot provide us with a description of the transition period itself. This transition, which is now upon us, will be assessed after we have experienced more fully the conflicts it creates.

I

Machine Design

1

Mechanization: The Assembly Line

The history of industrialization is evoked by three terms: the division of labor, specialization, and mechanization. In his great work *Mechanization Takes Command*, Sigfried Giedion shows how productivity increases with the division of labor; how the division of labor gives rise to specialized machines for performing particular actions; and how these machines, when linked progressively, mechanize the entire production process.[1] Adam Smith grasped the essential unity of these propositions two hundred years ago: "The division of labor is limited by the size of the market," he wrote—large-scale production accompanies specialization. The familiar features of the industrial period— the specialized machine, the mass market, and the disintegration of craft work—are all encompassed by this overarching conception.

Closer examination reveals a more complex situation. We see, for example, that not all machinery grew out of a prior division of manual labor. Some devices, such as the early textile machinery (the loom, the spinning jenny, the self-acting mule) and the early metal-machining tools, sprang from a mechanical imagination that could simulate human action without first analyzing it into components. The metalworking lathe evolved from woodworking tools, and improvements such as the slide rest (which permitted better control and easier feed of the tool to the piece being shaped) added new functions and elements to the machine rather than just copying the familiar movements of the worker or craftsman.[2] Yet the grand historic theme is essentially correct.

The Assembly Line

No single image of industrial technology expresses Adam Smith's conception of mechanization better than the assembly line. While assembly involves only 10 percent of all industrial labor, it nonetheless appears to be the culmination of the drive toward mechanization and supermachines, the development of continuous and controlled movement, and the extinction of all work-related skill. As Giedion says, "the assembly line forms the backbone of manufacture in our time." Four principles of technological design and organization have given the assembly line its mechanistic cast. Parts standardization, the continuity of production, constrained or rigid machinery, and the reduction of work to simple labor together characterize the prototypical mechanized production process and contribute to the evolution of a mass-production society. Standardized parts can serve a mass market. A mass market can in turn support specialized machinery, which must rigidly and reliably perform the same sequence of actions over and over again. Such machines, linked together in a continuous process, create a highly productive, highly specialized supermachine. Finally, workers are denied all exercise of skill and forced to perform simple and repetitive movements.

Standardization
Assembly work is so familiar to us that we take for granted what would astonish the machinist of a century ago. Workers can confidently attach screws, bolts, ball bearings, shields, and the like to a workpiece passing along a conveyor belt, never doubting that the parts will fit together. But the production of standardized parts that meet precise size and depth requirements has been achieved only after a long process of development in precision manufacture.

The first machinist to develop standards of precision on any significant scale was Joseph Whitworth, whose Manchester shop, set up in 1833, was among the first in England to sell machinery on the open market rather than for special order.[3] Over the course of thirty years he developed a series of ring and plug gauges that mechanics could use to measure the dimensions of their workpieces. But his gauges were too expensive for the average mechanic's shop and were suitable only for cylindrical work.[4] By the 1860s specialized gauges had been developed for all the parts of a particular item, such as the components of a hand revolver or a sewing machine. But the matching of gauges to parts could not keep up with the variety and range of materials produced. Later the more flexible "go–no go" gauges could set upper and lower limits for particular pieces, and the 1880s saw the spread of micrometers for fine precision measurement and production.[5]

The development of precision machining was critical to mechanization. First, the parts of a fully assembled product were now sure to fit together (although through much of the nineteenth century fitters were needed to size the roughly cut workpiece). Second, and more important, precise standards made possible the production of interchangeable parts.

Interchangeability is fundamental to the entire industrial apparatus. Products must be sold on a mass scale to mass markets, and the consumer cannot return to the point of manufacture for part replacement or maintenance. Rather, a separate parts economy must evolve so that the consumer (or the secondary producer) can confidently purchase a part that will fit the original product or machine. Mass marketing depends on interchangeability. Once parts are standardized, they can become the modules for the assembly stage of production. Thus, for example, a wide variety of guns and rifles can be produced from the same elementary parts combined in different ways. But this flexibility is possible only if the parts are manufactured to precise standards.

The principle of interchangeability of parts was first applied to arms production in the United States (though not, as myth would have it, in Eli Whitney's factory),[6] spreading to clock and watch production and then to the manufacture of belts, screws, and nuts. By the turn of this century standardized gauges and measures were available and accepted nationwide. The development of interchangeability depended upon the simultaneous development of machine-tool technology. Precise and automatic machinery could produce improved standard tools for making interchangeable parts. The process was self-feeding, since precisely machined parts could themselves be used to construct the precision machine tools.

Continuity

Assembly work also demonstrates the principle of continuity. Sigfried Giedion places the vision and the reality of continuity and continuous motion at the heart of the industrial revolution. Yet, as he himself notes, the term *assembly line* is modern. It first appears in the *Oxford English Dictionary* in 1930, earlier editions having described *assembly shops* or *areas*. The notion of a continuous moving line is apparently a novel one.[7]

It is true that many historians trace the concept, and the limited realization, of the assembly line back to the early nineteenth century. Early biscuit making for the English navy was based on the conveyor belt;[8] in the United States, hog processing methods in the 1870s inspired Henry Ford to devise his first assembly line, for constructing magnetos—only after this experiment did he develop an assembly line for the total Model T chassis.[9] Yet technical problems consumed industrial investment throughout much of the nineteenth century. Engineers could give little thought to integrating the flow and movement of materials through the assembly process when parts themselves were not machined to specification and machines were not supplied with steady power.

Let us consider hog processing—more precisely, the butchering of the carcass while it moved down an overhead conveyor—as an application of the assembly-line principle. The "parts" were organic, not metal; the hog was dismembered, not assembled; and no automatic tool could be devised to cut up its body. Thus production was entirely dependent on hand labor (aided, to be sure, by cutting tools), and only a systematic organization of that labor could increase productivity when other machinery could not.

Ford's invention of the assembly line to produce the Model T was thus truly—and self-consciously—revolutionary, for it brought the principles of conveyance and controlled movement to a metal-based industry where the problems of standardized parts and steady power had first to be solved. As one engineer wrote in 1915, "The assembly line is of historical importance, as being the first moving assembly placed in work anywhere, so far as is revealed by information to date."[10]

The principle of continuity places the single automatic or semi-automatic machine within a supermachine—the factory itself. The nineteenth-century factory remained essentially a job shop, with various machines placed randomly about in corners and on different floors, their individual motions controlled by a large wheel, often placed in the basement. Steam power was transmitted vertically through the factory building from the basement to the top floor: primary belting transferred motion to secondary shafts, which in turn transmitted power via pulleys to individual machines. The movement of materials, inventory, and work in progress through the factory was a haphazard affair.[11]

Factory layout became more rational only when electricity became the prime source of power. After the turn of the century the spread of electricity and of individual electric motors for separate machines freed machine placement from the constraints of belting, shafts, and the primary source of power. For the first time machines could be laid out according to the rational flow of materials through the factory. At the same time the owners of machine shops, inspired by the new doctrine of scientific management, were learning to devise routing slips, inventory tracking methods, and an entire range of techniques for organizing production: there arose a theory and practice of production engineering that encompassed the entire range of activities under the factory roof.[12]

These two developments, electricity and scientific management (the latter became known as industrial management after World War I), established the technical basis for the principle of continuity. Ford, though certainly a genius, was also a man of his time. The continuity of his assembly line had clearly been emerging from the interplay of technical and managerial development.

While assembly based on the moving conveyor belt was a particularly graphic example of continuity, depending as it did on a large corps of workers whose actions resembled more and more the motions of the surrounding machinery, the principle extended into other directions as well. Most striking was the production of automobile frames. The A. O. Smith Corporation of Milwaukee perfected a plant that turned out 420 frames per hour by first forming the frame with automatic machine tools and then riveting it automatically. The plant employed two hundred workers—one-sixth of its previous twelve hundred—and depended upon the automatic transfer of parts between machines and the use of machine tools that performed multiple cutting operations. The single machine was incorporated into a unified, continuous production process.

Similar developments took place elsewhere. Between 1925 and 1936 the principle of continuity reached the electric lamp industry.[13] Up to 1920 electric lamp production had been organized as a batch process: each operator, on finishing a batch of parts, put them into a container that was hand-carried or trucked to the next department. After 1920, however, layout was based on a rational grouping of equipment; the movement of materials within and between machines was automatically timed; lamps were redesigned to eliminate certain steps in fabrication; and effective devices for feeding work into machines were developed. These and other changes welded the disparate machines and material components of the electric lamp factory into a single supermachine. Indeed, much of what passed for automation in the great automation scare of the 1950s was simply the application of the continuity principle to an ever-increasing range of industries, from grain mills to cylinder block machining to the production of foam mattresses. The spread of automation was no more than the raising of mechanization to a higher level.

Constraint

The assembly line clearly demonstrates the principle of constrained movement. The conveyor belt imposes its own inexorable rhythm:

each action must follow the preceding one with precise timing. Otherwise, incompletely assembled parts move down the line, becoming increasingly imperfect because succeeding actions cannot be completed. Errors accumulate until the entire line must be stopped to restore the balance. In the movie *Modern Times* the image of Charlie Chaplin racing down the assembly line to keep up with his work captures the inflexibility of the assembly line. Machining lines impose equally rigid constraints. A drilling and boring machine must accept a cylinder and feed it to the next machine according to a precise schedule.

This precision of timing highlights the fundamental character of machinery. A machine becomes increasingly perfect as it limits motion and action to a specifically designated set of activities within a specifically designated period of time. Without constraint—with variance or slippage—individual parts will not be machined properly and individual machines will be poorly interconnected. The historian A. P. Usher writes, "The parts of a machine are more and more elaborately connected so that the possibility of any but the desired motion is progressively eliminated. Such a transformation results in the complete and continuous control of motion."[14]

As long as individual machines remained disconnected from factory organization and as long as the factory itself was disorderly, constraint at the level of the machine was offset by variability at the level of the shop. Many machine shops are still organized on this basis. Automatic machine tools are combined in a modular fashion to produce a particular part, but the machines themselves remain unconnected; instead, as in the shop of a century ago, materials are moved by hand and factory layout is haphazard. Productivity remains lower than it might be, but what the job shop loses in efficiency it gains in flexibility.

The supermachine is constructed at the level of the whole factory; constraint characterizes not merely the individual machine, but the system of machines as well. James Bright, whose ground-breaking study deserves frequent mention, examined the evolution of this systems constraint by studying the spread of full mechanization to numerous factories after World War II. The following examples are taken from his work.[15]

A production company introduced an automated line to produce its most popular oil seal. In this line, based on the modern principles of continuity and mechanization, connecting hoppers conveyed parts to individual automatic machines. The constraint principle intervened: when the demand for the oil seal fell unexpectedly, the automated line could not easily be adapted to the production of other oil seals. Similarly, engineers in a rubber mattress plant had difficulty adjusting the density of the foam rubber supplied to the automated line as production shifted from mattresses to automobile cushions.

Here and elsewhere, Bright concludes that continuous production imposed severe constraints on the ability to produce different products with different inputs. From the perspective of the fifties, one might have predicted that just as automatic machines could operate only with standardized parts, so the full mechanization of the supermachine could operate only with the standardization of demand and a long-term plan for coordinating buyers' wishes with sellers' capacities. The vision of full mechanization assumes a consumer culture in which the technologies of production impose larger and larger uniformities on social life.

The Reduction of Work to Simple Labor

Finally, the assembly line typifies the reduction of work to simple labor, robbed of all inherent interest or value. The worker, making a few turns of the wrist or sweeps of the arm, becomes a specialized machine, doomed to repeat these simple mechanical movements eight hours a day, five days a week, fifty weeks a year. Nothing seems more brutalizing, not because of felt pain or discomfort—for many studies show that workers do adjust to the rhythms of movement imposed by the line and may derive pleasure from the experience of continuity—but rather because experiences and potentials are lost forever as intelligent people are robbed of their ability to think, puzzle out, and discover.

This feature of the industrial process was evident from its inception; surely Marx's observations on textile production anticipated every criticism of industrial labor today. But it is in the rise of Taylorist engineering and principles at the turn of the century that the modern caricature—the worker as machine—truly finds its origin.

Frederick Winslow Taylor, a pioneer in the field of production engineering, introduced the study of human motion within a perspective emptied of psychological and physiological content. The worker's motivation was reduced to the desire for more pay, while the constraints on speed and movement imposed by the worker's body were reduced to a simple formulation of the balance between activity and rest. Only later would research demonstrate that the psychological content of work powerfully stimulates or inhibits productivity and that the human body, in obedience to its own gestalt principles, imposes constraints on the content of sequences of activity.[16]

Taylor, interestingly enough, did many of his time and motion studies in settings where tools rather than machines predominated.

His most popular text, *The Principles of Scientific Management*, examined pig-iron shoveling and loading, a process requiring only shovels and planks.[17] His well-known disciple Frank Gilbreth, who was not a production engineer, studied human motion in the abstract, first within labor-intensive work such as bricklaying and later within the broader sphere of sports and the arts.[18]

From his study of labor-intensive work, Taylor argued that scientific management would yield the "one best way" to utilize human beings in production. Machine-intensive work, however, imposed rhythms that greatly constrained human motion, thereby limiting the worker's options. Taylor's followers split into factions: Gilbreth continued to study human motion, while most other Taylorists, such as Carl Barth, began designing tool rooms and machine shops that forced the worker to submit to mechanical rhythms. In answer to Gilbreth's charge that Taylorism was being reduced to stopwatch calculations, Barth, a production engineer, replied,

[our] time study work . . . has been connected with all-around machine work in which the time of machining a piece of work has been more a matter of the study of the possibilities that are independent of the particular operator than of the operator himself. Our gains have consequently come from the respeeding and rebuilding of machines, the institution of improved tool rooms and so forth, than from the subsequent time studies made of manual operations involved. . . .[19]

With the rise of modern manufacturing the principles of human engineering—and human concerns generally—became increasingly irrelevant as the search for the one best way came to signify a drive for maximum speed, limited only by worker fatigue or resistance. This reductive quality of later Taylorism was perfectly expressed in the assembly line:

The instruction cards on which Taylor put so much value, Ford is able to discard. The conveyor belt, the traveling platform, the overhead rails take their place. These are the automatic instructions. . . . Motion analysis has become largely unnecessary, for the tasks of the assembly-line worker are reduced to a few manipulations. Taylor's stopwatch remains, measuring the time of operation to the fraction of a second."[20]

Conclusions

The assembly line illustrates the fundamental principles of mechanization: standardization, continuity, constraint, and the reduction of work to simple labor. Taken together, these principles form the core of an industrial culture.

We may argue, to be sure, that under different sociopolitical conditions industrial technology may develop differently or may affect workers differently. Adjustments to industrialism are found everywhere, including self-managing committees in Yugoslavia, work-redesigning experiments in the United States, and the continued and persistent existence of the unproductive job shop throughout much of the industrialized West. Nevertheless, industrial technology, whether it impresses itself within a socialist, capitalist, Western, or Eastern milieu, always gives its stamp to the character of work. Industrialization is a transcultural process that shapes social life in the same mold everywhere.

How can we escape from these principles to what Giedion calls a "new balance," in which variety of experience, breadth of conception, and depth of understanding are given their due within the world of work?[21] Are industrial principles reaching their own internal limits in the organization of production? What form is postindustrial technology taking? To answer these questions, we will show how production principles are being inverted: how machines are becoming more flexible as constraint is relaxed, how continuity of production increasingly derives from the uses of error rather than the perfection of movement, and how simple labor is being transformed into pattern-finding and problem-solving activity.

2

The Relaxation of Constraint

The ideal machine maintains unvarying connections between its parts while foreclosing all redundant or unwanted actions, in order that the same workpiece may be shaped over and over again according to plan. The machine must transmit power to its parts; it must transform circular motion into several permutations (horizontal, vertical, and irregular) to shape the workpiece appropriately; and it must control the speed and direction of movement not only of its own parts but also of the workpiece.[1]

Elegance is the key to good design: the machine must perform all necessary actions, and no unnecessary ones, with the fewest possible parts. Although a machine with more parts might be better able to do all the required tasks, it would cost more to build and maintain; moreover, the risk of parts slippage would increase. Generally in a good mechanical design the same part or series of parts simultaneously transmits power, transforms motion, and controls the speed and direction of movement, in this way minimizing the number of parts and preventing unwanted action.

Consider the following example. I want to construct a machine that moves a metal bar horizontally to the right. I can do this with a simple rack-and-pinion gear arrangement (see figure 1). The design is elementary but elegant: the small gear, the pinion, simultaneously transmits power from the large gear to the bar, transforms circular motion into horizontal motion, and controls the direction and speed of the bar's movement.

Imagine that I now wish to complicate my machine by making the bar move back and forth. In Rube Goldberg fashion I add a clutch; another set of gears, which rotate in the opposite direction; two springs, one attached to each end of the bar; and two stops, one at each end of the bar's journey. As the bar moves to the left, it tightens a spring

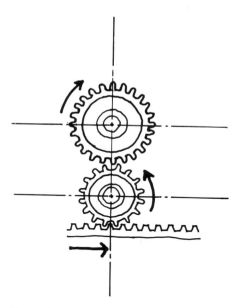

Figure 1
Rack-and-pinion mechanism.

that activates a clutch. This disconnects the small gear from the first set of gears and connects it to the second, reversing the direction of the bar. Moving to the right, the bar triggers the same sequence.

But like Rube Goldberg's machines, this is comical. I can bring about the necessary actions through a far simpler crankshaft arrangement in which a crank attached to a gear simultaneously transmits power to the bar, transforms circular into back-and-forth horizontal motion, and controls the length of the journey to left and right by its placement on the gear (see figure 2). The more automatic the machine and the more varied its actions, the more complicated its structure will be; but the fundamental design principle holds: a well-designed machine will integrate the transmission, transformation, and control systems within a minimum number of parts. This is always the challenge for the mechanical imagination.

An automatic spinning machine that combined the features of the spinning jenny and the spinning frame typified the magic that a mechanical imagination can produce. The self-acting mule was developed in England between 1825 and 1830 by an ingenious mechanic, Richard Roberts, a shoemaker's son and a student of the great English machinist Maudslay. The mule was organized around a carriage that shuttled

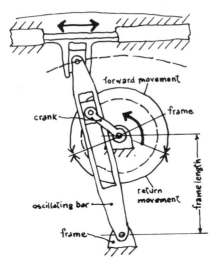

Figure 2
Crankshaft mechanism.

automatically back and forth along a track, pulling and twisting rovings, or loose thread, from an overhead bobbin onto a spindle that spun the thread into yarn. Observers marveled: Samuel Smiles, a biographer and self-help propagandist, called the mule "one of the most elaborate and beautiful pieces of machinery ever contrived";[2] and Marx wrote that the mule "opened up a new epoch in the automatic system."[3]

Yet such well-designed machines are also highly constrained ones, single-purpose in character and design and hard to modify. Consider the device depicted in figure 2: in order to change the length of the back-and-forth movement of the horizontal bar I must rebuild the crank-and-gear arrangement, placing the crank at a different point along the gear. In general, since the systems of transmission, transformation, and control share the same parts, modifying one system inevitably means modifying the others. If gearing is changed to transform motion differently, the transmission of power through the gears may be negatively affected: too complex a gear train will consume too much power. Because machines are mechanically organized by the interplay of masses, forces, torques, and linkages, their functions must narrow as they become increasingly automatic. In becoming more productive, they lose flexibility.

Machines can become flexible, then, only if they are organized on another basis. This is where the electric motor played a decisive historic role. Its advent did not merely change the source of power, so that the main shaft of a machine was driven by electricity rather than steam; electricity in fact modified the structure and design of machines, making them more flexible by progressively separating the transmission, transformation, and control systems without compromising reliability or cost. Beginning around 1900, the spread of the electric motor separated the transformation system from the transmission system; after 1920 the development and application of the vacuum tube (and, later, of solid state devices) separated the control system from the transformation system. The substitution of electrical for mechanical force in machine design created increasingly flexible and increasingly general-purpose machines.

The Electric Motor: Separating Transformation from Transmission

Faraday devised the first electric generator in 1831; machine tooling, however, remained unaffected by the development of electric power throughout the nineteenth century. The industrialist saw little advantage in simply disconnecting the main drive shaft, which powered all the machines in his factory, from its steam source and connecting it to an electrical source. The machine tools would still be encumbered by all the belts and pulleys that came off the main drive shaft. There would be no difference in either the placement of machinery within the factory or the coordination of parts within a machine.

The portable electric motor changed all that. The mechanic could now place two or more motors in a particular machine. The constraint on machine design was reduced, since different parts could move at different speeds without being connected to the same primary power source. Long gear trains were eliminated. In large machines, independent portable motors could now direct individual segments moving in different planes, eliminating the need for linkages that translated motions in one place to motions in another.

The use of electric motors to drive machines was first demonstrated in 1873.[4] At a Paris exhibition in 1881 electric motors drove sewing machines, lathes, drilling machines, and printing presses.[5] By 1887 fifteen well-known American manufacturers had produced over ten thousand motors of fifteen horsepower and below,[6] and buyers could use power from the Edison distribution system for electric lighting. By 1900 some advanced machine tools incorporated separate drive motors whose action was independent of the speed of rotation of the workpiece. "In 1914, 30 percent of American industry was electrified; in 1920, 70 percent was. Between 1919 and 1927, more than 40 percent of the steam engines in the United States went to the scrap heap."[7]

Example 1: The Paper-making Machine
Electric motors dramatically increased industrial production by improving overall factory layout and raising the speed of individual machines. Advances in paper making in the 1920s are representative here. John H. Lorant, in his study of capital improvements in the 1920s, writes,

Compared to water and steam power, electricity could be distributed through a plant far more easily, and power troubles in one department would not necessarily interfere with the functioning of another part of the plant. Furthermore, basements no longer had to be built to house prime movers, and the belting and shafting previously required to transmit power from these primitive engines to the machinery were no longer needed. Since electric motors transmitted power directly to the equipment, much of the cumbersome impediments required with steam and water power were eliminated.[8]

More important for our argument, the paper-making machine could run much faster. The Fourdrinier machine, developed at the beginning of the nineteenth century and even at that time largely automatic, was composed of parts that operated at different speeds.[9] The "wet" part, which suspends pulp in water, moves slower than the "dry" but unwieldy combination of gears, pulleys, shafts, ropes, and clutches that performs the rest of the operation. Eventually this cumbersome mechanical arrangement was replaced by separate electric motors— *sectional drive*—that independently ran the wet and dry ends of the machine at the proper speeds.

Electrical sectional drive had two advantages that led to increased machine speed. First, since many mechanical linkages were eliminated, there was less danger of mechanical slippage, or backlash, at higher speeds. Second, because the machine contained fewer parts, it consumed less operating power—that is, in transmitting motion along its parts it encountered less friction and slippage between them and could therefore commit more power to the direct transformation of pulp into paper.

Example 2: The Transfer Machine
Today, the complex "transfer machine," which integrates separate machine tools in the production of automobile engine blocks, can work only on the basis of sectional drive. A textbook on automation engineering explains that

as long as power was obtained from a water wheel, a steam engine, or from a large electrical motor that powered a number of devices, the problems of transmitting power to each mechanism and component of a machine required considerable ingenuity. The mechanical problems of transmitting mechanical energy where it was needed to all portions of the machine greatly complicated machine structure. Furthermore, mechanical components such as clutches, gear boxes, variable speed drives and linkages are not readily arranged to work together. Suppose a modern automation transfer machine had to receive all necessary power from a single overhead shaft by means of belting. . . . Problems encountered in the transmitting and interrupting of mechanical power to each stage of the transfer machine would make the operation of such a machine impractical. The trend now is to use a separate electric motor wherever power is needed. . . . Furthermore, the application engineering that is necessary in designing a motor drive can be done in a routine manner. That is not so when mechanical power transmission is used, for the development of linkages to transmit mechanical power involves specific problems that must be handled on a custom basis at each part of the machine.[10]

That electrification increased productivity is a common enough proposition. Our argument is more complex. Electrification increased productivity by reducing mechanical constraint within the machine. This permitted, on the one hand, greater speed, greater control, and smoother, more continuous production; and on the other, a more flexible machine that proved easier and less costly to modify. A simple example illustrates the point. To change the relative speeds of different machine sections, the mechanic or engineer, instead of stripping the machine and replacing old gears and cams with new ones, need only adjust the relative speeds of the different electric motors. As electrical forces were introduced into production, the pressing trade-off between productivity and specialization was relaxed. The relaxation of machine constraint opened the way to increasingly general-purpose machines, machines that could be modified at reasonable cost.

The Separation of Control from Transformation

The separation of movement control from motion transformation and power transmission created the technical basis for the flexible machine.
    Machine tools must shape metal according to plan and design. On the most elementary lathe the worker guides the cutting tool while the metal workpiece is turned or rotated on a spindle. Meeting the tool, the rotating workpiece is cut and formed by powerful mechanical forces. The worker, applying changing pressure to the rotating workpiece, shapes it to the desired size and pattern.

This design was much improved around the turn of the nineteenth century with Maudslay's invention of the slide rest, a device that sat on a long screw running along the base of the machine and held the tool. The tool could then be advanced along the screw toward the workpiece, and a single prime source or belt could drive both the spindle, which held the workpiece in a vise, and the tool feed. As Marx noted, the slide rest replaced not one particular hand tool "but the human hand itself."[11] This arrangement for automatically feeding the tool to the workpiece quickly gave rise to an elementary automatic screw machine. With a bar stock, or blank metal bar, fitted to its spindle, and with relative speeds set for the slide rest along the base screw and for the turning of the spindle, this machine could automatically cut a screw with a particular thread and pitch. Much as a simple modern key-copying machine uses one key as the model for another, the base screw became a template for the production of another screw.

Maudslay's lathe illustrates another point. From the beginning, automatic machine tools required some form of template so that the eyes and hands of the worker, which once guided the tool, could be replaced by a machine piece. The template became the control element of the machine, determining the feed rate and movement of the tool.

Cams as Control Devices

In the United States Thomas Blanchard developed just such a wood-working lathe in 1818, for shaping gunstocks at the Springfield Armory.[12] He used the cam, a fundamental building block of mechanism akin to the gear and clutch, to direct the action of the cutting tool. The cam is a specially—and often irregularly—shaped component "that serves to guide the motion of a follower attached to it."[13] The shape of the cam as it turns directs the movement of the follower (much as a rotating wheel can guide the movement of a piston), thus imparting its shape or some transformation of its shape to a tool linked to the follower. The tool will then shape a workpiece according to the information implicitly contained in the cam.

The cam, functioning as a blueprint, provides much greater flexibility in developing the form and timing of a particular motion than other linkage mechanisms, because a wide range of shapes and motions can be produced without stripping down and rebuilding the machine each time the workpiece must be shaped to a new design. Instead, one cam can simply be replaced by another. In effect, the control of movement is no longer contained in the array of gears, clutches, and mechanical stops that together make up the structure of the machine, but rather in the cam, which is separate from the body of the machine.

Blanchard's original cam-following machine was used only to shape wooden gunstocks. After the Civil War, however, metal-forming lathes based on cam-follower principles were introduced on a significant scale, providing a new level of flexibility for machine tools. Yet since specific cams were mounted as removable pieces on blank cylinders and could be replaced by new cams when the machinist so desired,[14] they imposed two critical limitations.

First, the cams themselves were made of metal and had to be shaped by hand. Shaping a cam to impart a particular motion to a tool can entail some complications. It is not easy, for example, to devise an irregularly shaped cam that will impart a sequence of horizontal and vertical motions to a milling cutter. The latter movements are trans-formations rather than specific visual copies of the cam shape, and geometric calculation becomes necessary.[15] Even today machine shops make only small quantities of such complicated parts; the fixed costs involved in producing the cam may not be recovered in selling the automatically produced pieces. In short, while the cam-following ma-chine is more flexible than an automatic machine tool, in which motions are controlled by mechanical linkages, its design produces rigidity and cost.

The second limitation is even more serious. The control system is separated from the power transformation system (the cam can be removed) but indissolubly linked to the transmission system. The cam provides both the guide for directing the tool and the force to move it. It is like a mechanical foot brake on an automobile: the pressure of the foot on the brake is both the signal to grip the wheels and the force to stop them. Only with automatic braking are these two systems disconnected: the movement of the foot becomes truly a signal that is amplified electrically into an appropriate force.

The linkage between the control and transformation systems of the machine places limits on the economics of cam making. Not only must the cam be powerful enough—that is, of sufficient metallic thickness and mass—to impart the necessary force; it must also be shaped so as to guide the tool accurately. Cams tend to be difficult and costly to shape and quick to wear out. Cam-following machines were thus used mostly for large runs that produced universal pieces such as screws and bolts.

The all-around machinist working on a general-purpose engine lathe remained indispensable in the production of limited-run parts ordered on a one-time (or infrequent) basis by local customers. While steam or electricity powered the machine, mechanics had to determine the proper feed rate, mix, and sequencing of the tools and the correct speed of the spindle. Much of Taylor's work in the 1880s and 1890s was devoted not to the study of efficient human movement but rather to the development of mathematical formulas for determining spindle speed and tool feed rates for machining a range of metal parts.[16] These constraints led shop owners to economize by splitting the workforce into skilled and semiskilled machinists. The former arranged the appropriate sequence of tools in the lathe's turret—a multiple tool holder that can present a succession of tools to the workpieces in a fixed time sequence—and the latter loaded and unloaded bar stock and finished pieces. This arrangement persists in many of today's shops. A more decisive separation of the control system from the transmission system, as in the automatic brake, would have to be built upon a different technological foundation.

3

The Technical Foundation of Control Systems

The Brain as a Model

The brain is a dense pack of millions of cells that interact through complex electrical and chemical signals to control the organs, cells, and chemical reactions of the body. Yet the brain uses relatively little energy in exercising control. The tennis player's arms, legs, lungs, and eyes consume far more calories than does his brain in controlling the coordination of these parts. This distinction between the low power consumed by the control system and the high power consumed by the actuating system is familiar in daily life. The flick of a switch requires a minuscule amount of energy, but it can unleash a cascade of BTUs as dynamos turn and electric power is produced at a generating plant. The energy supplied to the switch is amplified a thousandfold when the circuit is closed.

Viewed abstractly, control systems consume little power because they channel energy potential but do not supply any independent force. The switch, the valve, and the traffic light are essentially messages that open or close pathways of flow, providing a nonrandom organization to force and energy. The power consumed by these message systems—or, as we would call them today, communications systems—is only the power consumed in the making of messages. Communications systems are based on weak-power engineering, while mechanical, energizing, or acting systems are based on strong-power engineering.[1]

This distinction between weak- and strong-power systems is fundamental to the problem of machine flexibility. Imagine for a moment that the brain were organized on a different structural principle, that muscle tissue replaced nerve tissue in the cranium. Theoretically this is possible. Message networks can be organized with a wide range of materials: I can talk to you, I can write to you, I can hand-signal to you, I can do a complex dance for you which only you can interpret. The physical medium is irrelevant to the particular message; what counts is that both you and I understand the coding and decoding process and that an input message always translates to the correct output message, whatever the medium for translation.

So far, so good. But imagine how muscle-bound the brain would have to be and how much energy it would consume in controlling even the simplest of body movements. Before it could control anything, it would consume all the power available to it in simply moving the signal across its own networks. Such systems are not as fantastic as they sound. We encounter them every day in the form of large bureaucracies that expend most of their energy (budget, time) in maintaining themselves—in simply moving, channeling, and controlling messages through their internal communications systems. These bureaucracies are necessarily unresponsive to external demands; it seems inevitable that some new structural principle must emerge if they are to prove useful and helpful. Such bureaucracies do not *control* activities in their field or environment. Rather, by consuming so many resources they impede these activities.

Fortunately, nature has endowed the human brain with an immense capacity to handle the widest variety of messages within a medium and through principles of design (for example, the use of chemicals as message-transmitting agents) that require only the smallest amount of energy for their operations. This is a familiar proposition in comparative biology. We have a much more complicated and evolved brain than the ant's, because the ant has only a rigid repertoire of behaviors. Its brain need only consist of a set of fixed "circuits" that respond in predetermined ways to incoming messages and stimuli. In contrast, because of its dense and multiple modes of message transmission and its capacity to create message pathways, the human brain can respond to novel messages and, through learning, create a pathway or circuit within the brain and between the brain and body that corresponds to the novel message and its demands (that is, the learning when laid down as a circuit becomes a productive habit).

The reader must excuse this excursion into biology—or, more particularly, biology as metaphor. But the excursion is necessary. The development of flexible machine technology rests ultimately on a technological and structural design within which a control system of immense complexity can operate at low levels of power, can be modified independently of the transformation and transmission systems of the machine, and finally can be connected to the machine through some device or mechanism that translates its low-power messages into high-power actuation, such as cutting, shaping, or grinding. The history of manufacturing technology after 1910 is the history of the development of just such a system. Feedback theory and the vacuum tube are its bases.

The Concepts of Feedback and Learning

Imagine a carpenter building kitchen cabinets of soft white pine. We can watch her bang nails into the wood effortlessly, like a precision machine, so that with a few bangs and taps, the nail sits squarely, perfectly in the joint. If we examine her actions in greater detail we find that we can describe her nail-banging as a productive habit, a fixed set of rules specifying a sequence of machine-like motions: "Do one hard bang, one soft bang, two hard bangs, and then a tap." This sequence is invariant, with very minor exceptions, because the cabinet size, the wood thickness, and the depth of the joints are unchanging. From this perspective the carpenter is much like a machine following a rigid sequence of transformations from one state (bang number one) to another (bang number two), playing out in this way an effective procedure for getting the nail into the wood.

Imagine now that our carpenter must change over to hard yellow pine. If she follows the old sequence, she will crack the wood. Instead, she must alter her orientation. Each bang becomes an experiment. She taps and bangs the nails and then observes their depth of penetration and the state of the wood.

Her behavior has changed. She no longer follows a routine or set of rules; rather, she specifies a goal, takes an action, measures the difference between her goal and the consequences of her action, and takes another action. Were she now to work only with this hard wood, we could be confident that over time she would develop a new machine-like routine for banging nails into the hard wood. But if the wood changed daily, she would have to develop a different orientation toward her work. No "wood-specific" sequence could determine her work, since any habitual sequence would be useless with a new kind of wood. Instead, specific sequences would be replaced by the "metasequence," or sequence for determining other sequences, in which a goal is compared to some outcome and action is then taken to bring the next outcome closer to the goal. This metasequence describes the classic "feedback loop," as information about the consequences of an action in relation to a goal are fed back to affect the succeeding action.

This example illustrates two key points. First, as input conditions change, a feedback-based performance becomes the most effective way to achieve a fixed goal. Second, the role of error or difference is inverted as we move from machine or habitual sequences to feedback sequences. In a machine sequence, error or slippage undermines the effectiveness of the procedures. If the carpenter, habituated to hard wood, slips and fails to hit the nail on a particular bang, her motion is disrupted, and because she works from habit she must mobilize her conscious attention to correct her error. As we watch, her motion looks jerky and inefficient, for she is unable to adjust quickly to her error and find compensating motions to give her overall action a smooth and continuous look. In a feedback loop, error or difference plays a central role. The carpenter must import error into the loop because no known single action or sequence of actions will achieve the goal. Comparing the still-imperfect results of one action with her goal, she develops a succeeding action. If she uses the feedback loop to learn to adapt quickly to new wood—that is, if she learns to be a learner—she can develop a smooth sequence of compensating motions that appears efficient and continuous.

Athletes also experience this paradox of error compensation leading to continuity and efficiency. Focused on the ball and net, and experienced in the ways of his own body, the good tennis player continually adjusts body movement, arm movement, and head movement so that the stroke becomes smooth and efficient. Certain actions, such as the serve, may become habitual, but excessive habit generates rigid play. A mediocre player may master certain stroke sequences (hitting a lob into a powerful drive) but prove unequal to conditions that change second by second. The effective player is flexible, and flexibility is possible only if he uses the feedback loop as the basis for his action.

The feedback principle represents a shift in the concept and practice of transformation. Traditional machinery transforms by constraining movement until only perfect movement is realized. The resulting sequence is invariant. In contrast, the feedback loop transforms by importing error and developing a sequence of continually compensating movements so that a fixed outcome is achieved. The resulting sequence is flexible, changing as input conditions change.

Feedback Embodied

We have discussed feedback as a principle of transformation. How is it embodied in physical machinery? That is, how can the flexibility of feedback-based transformation and sensitivity to changing contextual conditions be incorporated into machinery to make that machinery similarly flexible? Imagine the following simplified example of a modern control device in a chemical production plant.

A photocell instrument measures and controls the density of mix between a liquid and a granular solid in a pipe (as in oil refining, when a catalyst must be added to a chemical cracking process). Light shining through a transparent portion of the pipe is sensed by a photocell, which translates radiant energy into a low-level electrical signal. The less dense the liquid, the more light shines through, creating a stronger signal that is then amplified by a transistor controlling an electrohydraulic valve. When the signal reaches a preset point of electrical strength, the valve opens and a very dense, pressurized mixture of solid and liquid is pumped into the pipe (rather than flowing around it). The density of the mixture in the pipe rises, the amount of light hitting the photocell falls, the signal from the photocell to the transistor weakens, and the valve closes. With the light source, the photocell, the transistor, and the valve preset at appropriate values (of electrical resistance, luminescence, etc.), the density level can be kept within narrow bounds.

The feedback loop is simple. The photocell senses the density and signals the transistor. The transistor amplifies the signal and sends it to the valve. The valve compares the signal with its preset level, or set point. When the signal is greater than the set point, the valve opens. Density rises as liquid is pumped through the valve, a new signal is fed back to the valve and, if the difference remains positive, the valve remains open. When, after a number of trips around the loop, density has risen sufficiently, the signal falls below the resistance level of the valve, and the valve closes.

This feedback circuit is based on the sequence of goal specification, difference measurement, and the reduction of difference that we described in the case of the carpenter. The control loop is structured so that any positive difference between observed and desired density is eliminated through action that increases density so that goal and actuality converge. The machine is flexible: it responds to a metasequence (the learning sequence of "compare, then act to eliminate differences"), rather than to a sequence based on particular conditions or inputs. There is consequently no one pattern of action such as we might find in a mechanical machine. There may be only one trip around the loop or there may be many, and with a somewhat more complex arrangement (two valves and more elaborate circuitry) the machine can act to decrease density as well. Such a machine responds to changes in its context.

The Special Role of the Sensing Device

The design of the feedback loop produces flexibility. In addition, it incorporates three physical features that distinguish it from mechanical machines. First, there is a separate and distinct sensing instrument—the photocell, for example. Second, there is an amplifying instrument that augments a signal from the sensing device so that mechanical work—such as the opening and closing of the valve—can be performed. Third, the amplification device channels the energy of the primary force of the system, in this case the pump. In a unified machine the transistor and the valve stand between the photocell and the pump, converting low-power messages into high-power changes.

This buffering of the message system from the energizing system is central to the effective operation of the feedback loop. Let us recall that feedback-based performance is based on the concept of continuous compensation. The tennis player's actions look smooth because he continually compensates for his head, hand, and body motions with small adjustments of posture and position so that he can meet the ball well. If he makes large, infrequent adjustments—swinging his arm too widely, moving his head too broadly—he overshoots his marks; he looks clumsy and his motions seem jerky.

Continuous compensation has two implications. First, constant vigilance is required. Because the sensing system must always be energized, more complex systems (for example, many photocells connected to circuitry that adds and multiplies their different signals) are economical and useful only if they use little power. Second, the sensing instrument must be able to detect small differences. A very sensitive instrument, however, will be unable to transmit mechanical energy or force directly. The addition of an amplifier produces signals with the power to perform mechanical work.

Imagine that instead of a photocell we insert a float (like the one used in toilet tanks) into the pipe and attach the float directly to the valve with a linking arm. As liquid density (but not volume) falls, the float will sink, pulling the arm downward and opening the valve. This feedback loop will be grossly insensitive, however. Density levels can change significantly, but until the weight of liquid displaced by the float drops enough to overcome friction and valve load, the float will not budge. When the weight of the displaced liquid reaches some critical number, the float may sink all the way to the bottom. Like the bad tennis player, this feedback loop will behave in a jerky and clumsy fashion.

The float functions so poorly because, as a mechanical device, it interacts directly with the conditions it is measuring. The valve opens and closes in response to the joint effect of the weight of the float, friction in the linkage, and the density of the liquid. A sensitive feedback system cannot use instruments that produce direct mechanical effects. The sensing instruments must not create or convey force; rather, they must have only sufficient power to sense and signal. Such devices will by definition use little power. We conclude that good sensing devices must interact minimally with the conditions they are measuring and thus must operate at low power. Consequently, if these signals are to prove useful, they must be amplified so that physical work can be done.

Feedback-based performance comprises a sequence of observations and actions in which discrepancies are detected and corrective actions taken. Such a sequence derives from the goal rather than the particular context in which the goal is being pursued. In contrast to mechanical performance, there is no predetermined sequence of actions; instead, the individual or the machine responds flexibly to environmental changes. In order to operate, feedback systems must import error, not exclude it as mechanical systems do. Finally, since effective feedback circuits are based on small, continuous compensations, the sensing device must be separated from the acting and energizing components of the machine. Feedback technology is therefore impossible without a concomitant technology of amplification. Since the vacuum tube was the first universal amplification device (we will examine this in the next chapter), we must conclude that the rise of feedback-based technology and flexible machinery was inseparable from the invention and development of the vacuum tube itself. Feedback technology is thus strictly a twentieth-century phenomenon.

The History of Feedback

Almost no text on feedback and cybernetics fails to point to the "governor" of James Watt's steam engine as an early example of feedback. In his device, "weighted arms [are] linked to the engine; as the engine turns at increasing speed, the weighted arms also turn faster. The arms, however, are mounted on pivots so that they are free to rise by centrifugal force as they revolve; the arms in turn operate a valve which admits power to the engine so that the valve is closed in proportion as the arms rise and the speed grows."[2] As the valve closes, less energy is supplied to the engine, which turns more slowly. The arms fall, the valve opens, and more power is supplied. Thus the speed of rotation of the engine is kept within narrow bounds, depending on the weight of the arms, the arm lengths, and the stickiness of the valves. The governor controls the speed of the engine on the basis of a feedback principle—or so it seems.

We may say that the weighted arms "compare" the actual speed of the engine to the desired speed (based on the mechanical linkages between engine arms and valve) and, "noting" a discrepancy, they open or shut the valve. But this is clearly metaphor. The governor does not, in any meaningful way, "sense": it is not a separate sensing instrument as is, for instance, a thermometer, a photocell, or an X-ray machine. Rather, it is a mechanical extension of the engine and valve, linked to the two so that feedback-like control emerges.

Otto Mayr, a thorough and creative historian of technology, examined Watt's own conception of his governor. Mayr notes first, "feedback devices described so far [until the invention of the steam engine] have remained obscure in unread books and in the shops of solitary inventors, or when serving on actual machines have played an inconspicuous role."[3] The device, he argues, "reached the consciousness of the engineering world" only with the invention of the Boulton-Watt steam engine. Yet, he notes, Watt did not patent the governor. In Watt's own words, "the invention of the centrifugal principle was not a new invention but had been applied by others to the regulation of water and windmills."[4] The governor, says Mayr, "did not strike Watt as new; he considered it merely an adaptation of a known invention to a new task. . . . One might infer from his silence that he did not see anything particularly significant in his principle. Compared with the large but straightforward task of producing power, the regulating devices and whatever theory they involved may have appeared to the sober Watt as secondary if not marginal."[5]

Indeed Mayr notes that texts and encyclopedias of the late eighteenth and early nineteenth century classify the regulator or governor as simply one among many mechanical controls similar in function to clutches, escapements, flywheels, ratchets, and brakes. A French text published in 1818 defines *regulator* as "a generic term given to devices whose purpose it is to regulate the movement of machines and to correct any irregularities in the motions."[6] Indeed, as Mayr finally notes, "In the time span covered in this study [up to the early nineteenth century] and indeed far beyond it, no author has been discovered who has referred to the closed loop mode of action as the common principle underlying a definite class of control devices. The inventors who design feedback mechanisms have employed the principle intuitively."[7]

In 1868, the brilliant physicist James Maxwell developed the first comprehensive theory of Watt's motion governor (though he was unable to present the full solution to his differential equations);[8] yet it is clear from his paper that he imagined the problem to concern the dynamics of motion, not the circuit of information. Restricted in this way, Maxwell's paper gave rise to the classical theory of speed regulation but not to a generic theory of feedback.

The historical verdict is clear. A society committed to the mobilization and transmission of power could only see control devices as constraints on machine motion. In principle and design, the governor was considered an extension of cams, clutches, screws, and belts. Moreover, as we have argued, only metaphorically could one describe the governor as a "sensor," the valve as a "comparator," and the steady-state speed of the engine as the machine's "goal." We are led to one inescapable conclusion: feedback-based control devices become important only when certain fundamental problems of power generation and transmission have been resolved and when sensing instruments are developed that are physically and conceptually distinct from their machine infrastructure. We are now ready to examine the critical historical role of the vacuum tube.

4

Feedback and the Vacuum Tube

The vacuum tube made it possible to apply feedback controls to machine systems. Its capacity for linking weak- and strong-power systems through a process of amplification enabled engineers to use electrical forces not only to supply power to machines, but to control them flexibly as well.

As one engineer notes, "I never cease to wonder at the fact that more fuss is not made about the invention in 1907 of the audion triode [vacuum tube] by De Forest. Everybody knows that Marconi invented radio, Edison the phonograph, the Wright brothers the aeroplane, but relatively few people know of De Forest, whose invention really has transformed our civilization. Such a simple thing to put a grid between the filament-cathode and the plate. And such a momentous result. Try to picture modern communications, including radio and television; physical sciences, such as nuclear science, physical chemistry, and solid-state physics; life sciences, such as biology and neurophysiology; or defense and space technology, navigation and guidance; and many other fields of human activity—without electronic amplification."[1] The vacuum tube, states another source, "is the symbol of an entirely new technology."[2]

Its Physical Properties

The vacuum tube is an electronic gate. A tube is emptied of air and a thin wire grid is placed between a tungsten filament, like that used in a light bulb, and a metallic strip. When electric current is applied to the filament (the cathode), electrons flow through the grid to the positively charged strip (the anode). Small voltage fluctuations applied to the grid are amplified in the current moving from the cathode to the anode. If the grid is supplied voltage or charge from the signals sent by a sensing unit, variations in the weak signal charge at the gate will be matched by proportional variations in the much stronger current that flows between filament and metal plate. This is the basis of electronic amplification.

The grid is like a venetian blind covering a window on a bright sunny day. A flick of the wrist can flood the room with sunlight and a similar flick can darken it completely. Small variations in the wrist flick result in amplified variations in the amount of light in the room. The sun's energy remains constant, but the amount that gets through depends on the position of the blinds. The variation of light in the room is an amplified version of variations in the wrist flick.

The vacuum tube originated not in manufacturing technology but in radio engineering. Its predecessor, the audion tube, invented three years before, rectified radio signals so that inaudible high-frequency signals were lowered in frequency at the receiving set. Later the vacuum tube was to amplify these signals so that a radio receiver produced strong signals of the appropriate frequency. Indeed, for the first twenty years of this century, the vacuum tube played its central role in radio, and improvements in its design and utilization emerged from engineering problems in radio design and transmission. As one engineer said in 1921,

The matter of greatest interest at the present time is the remarkable developments which have taken place in the thermionic valve [vacuum tube], both as generator, detector, and amplifier of electric oscillations. We are only at the very beginning of this evolution, yet it has already completely revolutionized the practical side of wireless telegraphy as well as telephony with and without wires. The whole planet is now converted into one vast auditorium . . . aided by the greatest electrical marvel of this half century, viz., the thermionic valve and repeater. . . .[3]

Amplification: A Communications Framework

Amplification may appear to be a rather simple technological concept. After all, the energy committed to the flick of a switch is amplified when the resulting closed circuitry unleashes large dynamos or generators. If I flick a switch rapidly, the on-off movement of the motors driven by the circuit will amplify the frequency of the on-off wrist movements.

The vacuum tube has, however, more subtle properties than does a simple switch. Central to its performance are its sensitivity and variation. Unlike the switch, which is constrained to an on-off amplification, the vacuum tube can amplify over a wide range of voltage inputs. Similarly, it can reproduce, at greatly amplified levels, high-frequency cycles in which variations in input voltage are transmitted by the millisecond. To understand its usefulness in manufacturing, consider the following problem. Hot steel ingots must be flattened and stretched for use as girders in construction. The ingots are pressed between large rollers controlled by powerful motors. As the motors exert greater pressure, the steel becomes flatter. The faster the ingots move between the rollers, the faster must the rollers change their pressure in response to changing conditions over the length of the ingot, such as the presence of impurities that cause variations in steel thickness.

Now imagine a feedback-based control device in which an X-ray gauge sends rays through the ingot. The thicker the ingot, the fewer the rays that penetrate and the weaker the signal reaching the detector on the other side of the ingot. The signals then pass from the detector through a series of vacuum tubes, and these amplified signals are further strengthened by a gas-filled electron tube. Finally the signal reaches the motor controlling the presses. If the incoming signal is below a certain strength, the top press moves down against the ingot; a stronger signal causes the press to move away. Thus the ingot is appropriately flattened along its length so that it emerges with some predetermined and uniform thickness.

This simple example illustrates the three essential attributes of the vacuum tube. First, the tube amplifies weak detector signals, which may then be used to drive large press motors. A signaling device of great sensitivity that runs on a low-power system can thus actuate a system of large power. Second, the tube allows the presses to modulate. No longer restricted to simple on-off movements, the top press can sensitively follow the surface of the steel ingot. Third, the tube transfers frequency response from the detection device to the motors, limited only by the top rate at which the motors can change their drive or speed.

Amplification, modulation, and *frequency response* are familiar terms, evoking the pervasiveness of twentieth-century communications systems in our lives. But the emergence of communications engineering and the modern communications industry do not simply reflect a shift in economic structure or the rise of a new market. On a more basic level, the machine has developed into a communications apparatus. The transmission of information, not power, has become its primary purpose. Effectiveness is measured no longer by the production or transmission of energy and power but rather by the sensitivity and accuracy with which changing information is transmitted to the power-channeling and actuating devices.

The machine as a communications apparatus is no mere conceit. Only through reinterpretation and reconstruction as a communications device can the machine play an effective role in feedback-based production. Norbert Wiener puts it well: "From one standpoint we may consider a machine a prime mover, a source of energy . . . a machine for us is a device for converting incoming messages into outgoing ones."[4]

Moreover, the complexity of the control apparatus can be increased without a measurable increase in the power consumption of the machine as a whole. Thus, for example, a second X-ray machine can be added that detects thickness at a different cross section of the ingot. The resulting signals can then be averaged by a simple electrical circuit (properly organized circuits can perform the arithmetic functions) and the average signal amplified and sent on to the motors. With more complexity (such as an impurity detector based on a photocell that detects reflected patterns of light and shadow or a circuit that correlates impurity presence with slab thickness to create a signal that weights the two indicators of variance in the steel), the control apparatus, like a brain, combines detection, computation, and message-transmitting capabilities within a low-power system of great flexibility. The relaxation of machine constraint takes on real meaning as the control system assumes a separate, distinct physical existence. The machine becomes a communications apparatus.

Feedback and the Vacuum Tube Together

I have discussed the feedback principle and the vacuum tube as two distinct, unrelated elements in the modern communications machine. While this is conceptually correct, the two must be understood as practically interdependent; indeed, engineers first formulated the theory of feedback in applying vacuum tubes to communications systems.

The design of sensitive feedback circuits that effectively compensate for outside disturbances is a difficult undertaking, though some of its aspects are easy to understand. Suppose you want to take a shower. You turn on the cold water tap, wait a second, then turn on the hot water tap—but the water remains cold. Turning the hot tap some more, you find the water is still too cold, so you increase the flow and soon the water becomes scalding hot. You quickly turn back the hot tap, but the water is still too hot, so you turn it farther back. Still too hot. Turn it once again—but now the water is too cold. Somehow you can never get the water temperature just right. Why not?—because it takes time for the water to respond to the instruction you give to the tap. When you don't wait—don't lag your response to your detection actions—your response is always ahead of the action of the water system; consequently, you are always overcompensating. The system as a whole—you, the tap, and the water—hunts for some never-to-be-achieved ideal water temperature.

This represents the problem the early designers of feedback devices faced. They needed a theory that specified the interactions among lags, frequencies, amplitudes, and information loops. The solution arose, however, from an entirely different context, as is often the case in technological breakthrough periods in which disparate fields of inquiry interact and interpenetrate.

Telephone and Radio Engineering
Radio engineers faced a dilemma. If they amplified radio signals by using tubes in series, the transmitted signals were greatly distorted, particularly in long-distance calling. Even with many multichannel amplifiers connected in tandem, the signal was often distorted. For years engineers sought to improve tube characteristics with no success.

The solution of H. S. Black, a Bell Telephone engineer, was inspired by the engineer Charles Steinmetz, who emphasized that every problem must be attacked in terms of its fundamentals.[5] In a major conceptual leap, Black conceived of his problem operationally, or functionally, rather than in terms of tube characteristics. He saw that he had to remove distortion from a signal rather than produce a perfect signal. Instead of trying to perfect the tube, he began to design circuits of tubes in which the movement around the circuit would ultimately produce an undistorted signal. After four years of work he hit upon the correct and most simple formulation. The signal plus its distortion had only to be fed back to its input to produce an undistorted signal. In a message transmission system with particular frequencies of transmission and a defined "lag structure," feedback could regulate and stabilize overall performance in spite of outside disturbances and "noise." Error was imported and then overcome rather than eliminated by design.

Physicists and engineers quickly established the precise mathematics of feedback structures for radio circuits in particular and for control devices in general (then called servomechanisms). By 1930 the fundamentals had been established; now the long work of developing a theory of control and communication began. The Depression intervened, and theory could meet practice only when the demands of war and the design of weapons systems placed great pressure on engineers, mathematicians, and physicists to develop tracking radar and effective antiaircraft guns. Yet there is little doubt that the revolution in cybernetic thought began with Black's discovery. One Bell scientist expressed it this way: "Although many of Black's inventions have made great impact, that of the negative feedback amplifier is indeed the most outstanding. It easily ranks with De Forest's invention of the audion [vacuum tube] as one of the two inventions of broadest scope and significance in electronics and communications of the past fifty years."[6]

The Cybernetic Framing of Problems

Black's conception of the distortion problem—his shift from individual tube characteristics to the relationship between tubes—is a quintessentially cybernetic way of thinking. In a mechanistic world view a machine's performance is only as good as its parts. In a cybernetic framing the parts and their valence, function, and performance are seen in relation to a whole, in the way they are embedded in a circuit.

There is evidence that the cybernetic framing of problems was emerging generally among scientists during the 1920s. In 1924 the German psychologist Köhler in his book on physical gestalts formulated what could be called a general systems theory.[7] The biologist Lotka published the first work on open systems; and Cannon's *The Wisdom of the Body*, published in 1932, developed the concept of homeostasis.[8] These studies are themselves rooted in Bergson's turn-of-the-century attempt to develop a philosophy of the organic and the vital that could stand as distinct from analytic science based on inorganic realities.[9] Common to all these attempts was an effort to discover and describe dynamic aggregates in which the whole is not only greater than the sum of its parts but also has self-preserving and self-sustaining properties. Undoubtedly the concept of an electric circuit, in which the substantive reality was small—thin wires (in radio, no wires) and invisible electricity—but the capacity for action and organization were great, gave impetus to these efforts.

The Vacuum Tube and the Steam Engine

Just as the steam engine enabled engineers to tap stored energy, so the vacuum tube enabled them to extract, organize, and use information. Moreover, both the steam engine and the vacuum tube were developed in the context of qualitatively new ideas about the physical universe.

The steam engine design was based on Galilean and Newtonian concepts of force and work.[10] The Copernican model of the solar system and the Newtonian conception of absolute space ended all speculation that cosmic space might be filled with air, which because of the earth's rotation would blow irresistibly from the east. If the atmosphere was finite and local to the earth, then, like a finite ocean of water, it exerted downward pressure. This led very quickly to the notion that air pressure, like water pressure, could be harnessed to do work. The steam engine, turning water into steam and then condensing it to create a vacuum in the cylinder, did just that: it employed air pressure operating against an air vacuum in a cylinder to move a piston on its downward stroke.

The steam engine also typified the new science of machine mechanics. Watt, a brilliant designer of machine mechanisms, not only devised the fundamental solution for the steam engine itself—a separate condensation chamber—but also invented key mechanical linkage mechanisms—parallel motion links, sun and planet gears, and the centrifugal governor—which set the tone for machine design as the practical application of the science of forces and motions.[11]

Similarly, the vacuum tube was designed only after inventors and scientists discovered the electron and understood some of its properties. This discovery opened physics to a new non-Newtonian scheme in which waves could be described as particles and could be propagated invisibly through empty space. From the mystery of wave propagation (if waves moved through space, what was the medium of their propagation?) arose Einstein's theory of relativity. The technical forerunners of the vacuum tube played critical roles in clarifying the nature of the electron. Roentgen's X-ray machine was an electric light from which most of the gas had been exhausted. J. J. Thompson discovered electrons by using X rays. Finally, building upon Thompson's work and that of another physicist-inventor, Fleming, De Forest was able to develop the amplifying potential of the tube itself.

Thus, the vacuum tube and the steam engine played similar roles in the history of technology. One tapped energy; the other, information. One was based on a Newtonian world view; the other emerged from a non-Newtonian world view. One gave great impetus to the technology of mechanical linkages; the other, to the theory of cybernetic feedback. The invention and use of the vacuum tube signified that technology was being organized on a new, nonmechanical basis.

Flexible Manufacturing: Two Sectors

Together, the vacuum tube and feedback theory enabled engineers to separate the control and transformation systems of machines. The emergence of separate control systems altered the fundamental assumptions and consequences of machine design. Rigidity, constraint, and specificity, the three principles of industrial machine design, gave way to flexibility based on general-purpose, uncommitted machine systems. The history of two industrial sectors, petrochemical production and machine tool manufacture, exhibits this shift to a new principle of machine design.

Chemical Manufacturing as the First Cybernetic Industry

Giedion describes the era between the two world wars as "a time of full mechanization";[1] in the case of most metalworking and fabrication industries he is correct. Developments in automobile production are most striking here. The transfer machine for linking machine tools, synchronous scheduling systems for meeting rigid timetables, and rationalized assembly processes together constituted a paradigm of mechanical production systems. But in the quintessential twentieth-century industry, chemical and petrochemical production, we encounter a different situation. In the first place, chemical production is an inherently nonmechanical process, based on flows, volumes, and invisible molecular reactions. Moreover, as engineers integrated the different steps in the chemical production process, they developed the first coherent set of feedback-based control devices. To be sure, when they began (in the decade 1910–1920), control engineering was far from a science and feedback theory had not yet been developed. But by 1930, when feedback theory had been formulated and integrated chemical production facilities were operating, the convergence of vacuum-tube technology, feedback theory, and integrated production created the first cybernetic industry. Thus, the era of full mechanization also witnessed the first step beyond mechanization as the basis for production technology.

The Rise of Petroleum Refining

We tend to think of petrochemical production as the prototypical science-based industry. Yet petroleum refining, the production of gasoline, and the subsequent production of a myriad of chemicals from oil—alcohols, vinyl chloride, styrene, polyglycols, and the like—emerged as a viable and modern industry only after World War I. The proximate circumstances are obvious: the production of automobiles on a mass scale increased the demand for gasoline, motor oil, and various paints.[2] Petroleum was no longer used to produce heavy motor oils or lighting oils alone; instead it became a major element in the industrial configuration that was to propel economic development for some fifty years to come.

Before World War I petroleum refining was practiced pragmatically. No chemical engineering profession existed until 1908, the year the first chemical engineering textbook was published.[3] In the first twenty years of this century the typical petroleum refiner, "sensitive to the most obvious operating cost but little else, was responsive to experiments conducted along the simplest mechanical lines at insignificant low cost."[4]

Refining up to that time "consisted of simple distillation in which the oil was heated and the various vaporized fractions separately condensed. Refining was mere separation rather than conversion, with little flexibility in tailoring the output either to demand or to the highest-value products."[5] But under pressure to produce gasoline, refineries could no longer just separate the different components of oil on the basis of weight, using only the lighter parts for gasoline. The heavier parts had to be converted through intense, pressurized heating—"cracking"—through which large molecular chains of carbon were broken down into smaller ones. Refining and processing were to become a process of chemical conversion rather than physical separation.

The pressures of these demands led to many innovations, all of which placed refining on a continuous-process production basis. Three are representative. First, the intense heat and pressure applied to the oil did much damage to the oil shell stills, in which large batches of oil were heated like "water in a tea kettle."[6] Consequently these stills were replaced by tubular heaters in which oil was passed continuously and rapidly through pipes in a furnace. Since the volume of oil in the tubes was smaller at any given moment than that in the shell stills, refiners could apply much higher pressures and temperatures without damaging the tube walls. Moreover, refiners could also vary temperatures within a wider range, thus enabling the tube stills to handle a wider variety and quality of fuel stock.

Second, since the heavier components had to be cracked after distillation had removed the lighter ones, it made economic sense to run the oil through the cracking stills while keeping the temperature constant. Under the old batch process the crude would be run through the still, the light fractions "topped off," and the oil cooled down and then rerun through the stills. As the demand for gasoline rose, however, and the heavy fractions had to be converted, heating costs rose. This led to the development of fractionating towers, which redistilled the lighter components through a tubular heating process and ran the residual oil "straight to the cracking tubes, thus eliminating equipment, labor, heat loss, and duplication of work."[7]

Third, the introduction of chemical catalysts sped up the cracking process, reducing the required temperature and pressure. Catalytic cracking reduced costs while increasing quality and yield. Since the catalysts were not consumed in the cracking process, they could be recirculated to the reaction chamber for further use. As with the heavier crude components, however, any lag between uses of the catalyst would force the refiner to cool down the catalyst and then reheat it as it entered the chamber. The catalyst was therefore recirculated continuously from its regeneration or recovery point to the reaction chamber and back.

Continuous Process and Feedback Control

What has continuous process to do with feedback control? We have already argued that the industrial principle of continuity extends the power of one machine to a system of machines. Our example, the assembly line, expresses this industrial ideal of perfectly coordinated movement "down the line." The principle of continuity in petroleum refining, however, reveals something new and different. As discrete processes were linked together, each process became more vulnerable to potentially disruptive conditions. Separate processes, such as distillation, cracking, and fractionation, were not protected from each other. Thus, for example, the temperature of the reaction chamber was affected by temperature and amount of the recycled catalyst. Any variation in the latter, due to random chemical processes, changes in the quality of the catalyst, or changes in the rate of catalyst regeneration, would affect the temperature of the reaction chamber. With a buffer between the two, the catalyst temperature could be precisely controlled to match the requirements of the reaction chamber.

Continuity required that systems of continuously operating controls be developed and installed at the interstices of unit processes, at chemical barriers, and at energy transformation points. Such controls had to function at the same rate as the chemical and separation processes themselves. There would be no place for the human eye, hand, or nose here.

In the 1920s the petroleum refining industry, and the chemical industry in general, became the first to apply automatic controls systematically and consciously. The movement from mechanical to informational principles of transformation took concrete form. Integration of processes toward continuity cannot lead to perfect movement, however. Because integration increases the sources of error, the system as a whole must include controls that continuously compensate for these errors.

The controls used in the twenties were simple in design and concept. For example, flotation, temperature, pressure, and volume regulators all worked on simple pneumatic or hydraulic principles (most were not electrical because of the danger of fire in many parts of the refinery plant). Their sensing devices were nothing more complex than mercury thermometers, tachometers (governors linked to springs), thermostats, tube constrictors for flow measurement (Venturis), or float balls like those found in toilet tanks for volume regulation. The signals from these devices were in turn amplified by power systems from hydraulic or pneumatic pumps.

Based as they were on mechanical designs and principles, such controls do not fully meet the criteria for feedback devices discussed in the previous chapter. But because they used little power (springs, mercury tubes), they were relatively sensitive, and their signals were amplified by a combination of valves and pumps. Most important, however, they constituted the first autonomous system of controls, independent of the pumping of primary energy and the materials transformation system. The control system was finally separated from the rest of the machine.

Engineering developments reflect this change as well. In 1922 the American Chemical Society held a symposium on automatic process control. *Chemical and Metallurgical Engineering* reported at least twenty-six improvements in control devices during 1924; by 1929 the journal was able to claim that "just as continuous processing is everywhere replacing batch handling as soon as it can be applied profitably, automatic control is taking the place of manual operation as rapidly as it proves itself.'"[8] Indeed, a review of the *Journal of the American Institute of Chemical Engineers* from 1920 to 1932 reveals a growing awareness of the centrality of control to continuous-process production and of the role that vacuum-tube technology could play.[9]

These changes in petroleum refining and, later, in petrochemical production were repeated in chemical production in general during the decade. Specific process innovations (such as continuous cracking) played a smaller role, but the same general principle of development — continuous processing based on automatic controls — proved central and dominant.

Output Flexibility

The shift in design emphasis from perfected movement based on mechanical linkage to imperfect, error-based linkage using information and control had other implications for production. First, as error-controlled integration based on feedback developed, the level of flexibility and adaptability of the production system increased.

Because chemical production recombines the building blocks of materials, a particular chemical input can be transformed into a range of outputs as long as the determining conditions (temperature, pressure, volume, speed, catalyst) can be changed appropriately. For example, both cracked gas and cracked ethane can produce ethylene. Ethylene can in turn be transformed into vinyl acetate, ethylbenzene, and vinyl chloride with the appropriate additions of chemical compounds and catalysts.[10] Similarly, at least four different processes can make phenol from benzene, and about ten processes can make acetic acid, starting with three raw materials.[11] The molecules create entirely new materials with different properties. In contrast, the transformation of metals by mechanical means takes place at the visible physical level: chemical bonds remain intact, and much energy is required to reshape the materials despite the structure and resistance imposed by their underlying molecular structure.

Each transformation requires a different input of chemicals, catalyst, and operating conditions, inputs that cannot be built into the chemical processing equipment. A simple volume regulator, such as the float-flapper-valve mechanism, can be set at different points so that it stabilizes the volume of liquid at any desired level. Operating as part of an independent control system energized by a secondary low-power system (a spring), it can be modified quickly and cheaply without reorganizing the links or structure of the main piping, tubing, and heating elements.

We turn again to our fundamental feedback principle. Once we import error into a system of transformation, the system itself becomes more flexible. We can change the goal (in this case the operating conditions controlled by the automatics) without at the same time changing the means (the effective linkages). Finally, the versatility of chemical production extends to the units themselves, making it possible, for example, to convert a gasoline plant in a relatively short time into a plant producing intermediates used in making alcohol for plasticizers. Only the basic units of production—the fractionating towers and the filtration and evaporation units—remain constant.

Information Capital
The introduction of continuous process on the basis of automatic controls brought another fundamental shift in production. For the first time productivity increased significantly on the basis of capital-saving technology. Lorant, whom we have used extensively here, has made this fact the center of his own investigations. The data are striking. Lorant estimates that capital productivity in continuous thermal cracking increased 3.6 percent per year over the decade 1919–1929, or 42 percent for the entire decade. In other words, if one hundred dollars of investment yielded ten dollars of refined output in 1919 for a 10 percent return, the same investment gave fourteen dollars in 1929 for a 14 percent return.[12]

Similar trends in capital productivity occurred in general chemical production. The economic implications of capital-saving technologies are complex. From a macroeconomic point of view it is clear that after 1920 the entire manufacturing sector was transformed in this direction as the capital-to-output ratio fell for the first time in American economic history and capital itself became more productive.[13] The growth of automatic controls was linked to the increase in capital productivity: indeed, the former came to substitute for the latter. With continuous production, petrochemical plants become smaller, more compact, and more flexible. Capacity increased without the addition of fixed equipment, such as buildings, holding tanks, or moving equipment.

Both mechanization and cybernation can reduce capital-to-output ratios, since each integrates processes and eliminates much holding-in-process equipment, such as storage tanks and stacking cranes. The two are related developmentally. Mechanization is limited by its own rigidity: parallel machine systems must be maintained for different products. Here cybernation takes over, producing greater flexibility within the same machine. The growth of flexible as well as productive transformation systems means that physical capital—machines and hardware in place—is replaced by information capital—smaller, less visible control systems that use less energy, yet give tremendous leverage and flexibility to the physical equipment.

In sum, petroleum refining and the production of petrochemicals gave rise to the first cybernetic production process. It is clear that the period 1910–1930 was the turning point of postindustrial development. During those two decades vacuum tube technology was developed, the feedback principle was elaborated, continuous cybernetic production emerged in the modern chemicals sector, and information capital began to replace physical capital in many manufacturing sectors. The old elements of industrial technology—physical capital, mechanical linkage, and error-free transformation—were giving way to new combinations.

Machine Tools

In my survey of machine tools in chapter 2, I argued that the separation of the control system from the transformation system was limited by a machine design that made the cam the driving force for cutting the workpiece. Here, as in chemical production, new technologies emerged in the years between the world wars.

The first technical breakthrough occurred in 1921. John Shaw, the inventor of the Keller duplicating machine,[14] replaced the metal model with one of wood or plaster of paris. A tracing cam moving lightly over the model's surface changed the currents that were transmitted to a motor controlling the feed rate of the tool. This machine, used primarily for profiling large dies used in automobile production, embodied the technical principle of separating the control system from the energy-transmitting and materials-transforming systems. But there is little evidence to indicate that this machine had a decisive impact on machine-shop practice. The older automatics, turret lathes, and general-purpose engine lathes predominated until after World War II.[15]

In 1952 the Air Force and the Massachusetts Institute of Technology together produced the first numerical-control machine tool,[16] using the feedback principle and the automatic player piano. In this device, an electrical contact mechanism sensed perforated patterns in a tape much like a videotape. Amplified, the resulting signals activated the table holding the workpiece and the cross slide (which moved perpendicularly to the length of the machine tool) or the drill head. Because the tool could move in three dimensions, very complex instructions could be coded using this system. This was the player-piano principle. If the table moved too little or too much, an electrical signal would notify the motor to readjust the position of the table.

This elementary automatic machine tool demonstrated the two dimensions of a flexible design: a coding system that controlled the sequence of machine actions and could be modified without changing the structure of the machine, and a feedback system that helped create a metal shape independent of the machine's error-producing context. The machine tools developed since 1952 have been elaborations of the original MIT design. Engineers have provided new principles of movement (such as point-to-point versus continuous-contour cutting), more elaborate feedback devices, and tool-changing as well as tool-moving programs (the machine can automatically replace one tool with another as different cutting styles are required); computers have replaced simple tape mechanisms; and, in the most advanced machines, tool designs drawn directly on a computer screen can be automatically converted within the computer to a program of instructions for cutting the workpiece.

Flexible Manufacturing Systems

Just as numerical control has increased the productivity and flexibility of individual machine tools, so engineers in the last decade have created flexible manufacturing systems by connecting such tools with automatic handling, loading, and transfer machines. One firm uses "six general-purpose machining centers for milling, drilling, and reaming, and four specialized head indexers for drilling and tapping, all arrayed over a floor area of nine thousand square meters. The machines and several manned loading stations are joined together by twelve tow chains, which provide power to a number of four-wheel carts. These carts transport the individual palletized castings to machines in the order necessary for their processing. The workpieces are moved onto the machines by shuttle carriages, where a second computer initiates the direct numerical control transmission which causes the machine to process the part. After processing is completed, the pallet is moved off the machine and routed to its next work station or returned to a loading station for manual removal of the finished part. This process is carried on simultaneously for several dozen parts of various types which may be in the system at one time, routed in random order among the work stations."[17]

These new devices have significantly changed the economics of batch production of metal parts. In the past, faced with a diverse market, small machine shops had only two mutually exclusive options.

With special-purpose machine tools—the standard, the lathe, the milling machine,—they could manufacture a particular part at an annual volume of two hundred per year.[18] Alternatively, the smaller shops could invest in general-purpose machines, leaving specialized production to the larger firms. The efficiency of general shops is low, however. New tools, jigs, or fixtures are required each time a new batch of parts is produced, while the transfer of semifinished parts between different general-purpose machines creates a large stock of parts in various stages of manufacture. A particular metal piece (a specialized gear or a valve for a compressor) might be worked on only 5 percent of the total time it spent in the shop. The new, machine-integrated system can significantly reduce idle time by changing and routing the machine tools automatically.

Nonetheless, these automatic systems have significant technical deficiencies. Because far fewer workers operate the machine (reducing labor costs as much as 80 percent), there are fewer hands and eyes to monitor incoming materials; and because the process is continuous, there is less time to do so. Furthermore, the technology of sensing and correcting errors automatically is still underdeveloped; therefore, input specifications must be tighter. Even so, shopworkers often have to stop the machines to check the quality of the semifinished workpiece.

Although these flexible systems have not yet been perfected, it is easy to predict that in the essential elements of their structure and performance they will differ significantly from automated cutting systems based on mechanical principles. In mechanical systems, tight schedules and fixed routines make the production process rigid and unchanging. Giedion's description of an automobile frame plant is illustrative:

A completed frame leaves the conveyor end, brushed and clean for the paint line, every ten seconds of the production shift. It takes ninety minutes from the strip of steel as received from the mill to the delivery of an enameled automobile frame into storage. The material is worked upon and moves back and forth through the factory on the most varied types of conveyor systems in an uninterrupted process. First, in a subassembly line, often in parallel operations, the steel bars are cut, punched, and formed. A second group of machines assembles the various parts until they are finally clamped together in the general assembly line. . . . No longer is the individual machine alone automatized, as is usual in bulk manufacture. Here, extremely precise time charts guide the automatic cooperation of instruments which, like the atom or a planetary system, consist of separate units, yet gravitate about one another in obedience to their inherent laws.[19]

In contrast, flexible, computer-aided systems will be able to route the same parts through varying sequences of machinery; accommodating rapid changes in part design, they will be able to produce mixtures of loosely related parts at the same time. Through an evolving set of controls and sensors, workers will coordinate and inspect materials and will monitor the entire system. Automation in the chemicals, cement, and plastics industries has already demonstrated the flexibility that control systems bring to machines. But the emerging machine systems will demonstrate an even higher level of flexibility as small batch production of parts is organized through a continuous process.

The Systems Impact

Such machine systems affect not only production planning but also cost accounting, investment planning, and strategic planning. With a 50 to 80 percent drop in labor costs, accountants must calculate the value of machine hours, which depend on assumed rates of depreciation.[20] Depreciation rates are not objective figures, however; they depend on the firm's investment policy—the rate at which it decides to replace machinery—and on the subjective rate of discount of future costs. The present value of a machine will depend on its purchase price, its estimated length of use, and the rate at which the firm discounts future costs to the present. (The latter may be only loosely tied to current interest rates.) Similarly, to measure the investment value of a particular machine system, cost accountants must measure the value of machine flexibility or potential.

To evaluate an investment in a special-purpose machine, a financial planner can examine the historical record of prices for the machine's product. But because a general-purpose machine system, which can cost up to ten million dollars, may have unforeseen uses, planners cannot easily calculate its potential. Each of the many dimensions of flexibility conveys a particular economic advantage to the firm deploying the technology.[21] The shop using the machine system can produce different mixes of parts at the same time, changing the composition of the mixes at will. Parts can travel between machines without rigid space and time constraints. Such a shop can respond quickly to changes in design or in the volume of output. All these factors contribute to the economic value of the machine system. Ultimately, the decision to invest in a particular system will depend upon the owners' strategy for the firm's development. Does the firm most need design-change flexibility, volume flexibility, or parts flexibility, and in what combinations? The flexible machinery thus integrates the once-separate departments of production planning, cost accounting, financial planning, and strategic planning. Because the machine system's economic value includes the potential for change, it must be measured against the overall strategic posture of the firm.

6

Control or Learning?

Since the beginning of the twentieth century, job designers, industrial engineers, and efficiency experts have tried to synthesize the informal, tacit knowledge of craftsmen—with only partial success. But modern control theory and the computer promise to bring this goal within reach. Through World War II most control systems were electro-mechanical devices that translated amplified signals into physical quantities, such as electrical resistance, water level, or air pressure; they were known as analogue devices. Like all physical devices they were subject to wear and tear, they imposed limits on the speed of information transmission, and they could not store past measures in a long-term memory. Most important, the system of controls could not be represented in a single unit or device. Although it is possible to design electrical configurations that represent a particular set of mathematical relationships—so that analogue simulators could theoretically have integrated the separate signals within a complex circuit modeled on the production system—such devices would have proved unwieldy and energy-inefficient.

Beginning in the early 1960s, however, computer technology wedded to solid-state circuitry provided a new technical basis for controlling production. The new devices were smaller and their electrical output could be fed into a digital computer. In contrast to an analogue sim-ulator, the digital machine could integrate large amounts of information without wasting energy or using complex circuits. The rules for in-tegrating the data were embedded in a program, not in the hardware of the circuitry itself. Now engineers had an instrument for integrating and controlling the controls. Data from all over the production process could be fed into the computer and integrated into the software model of the production process. Computer-generated instructions for feed-back control could be amplified by the transistors. Nonetheless, engineers attempting to analyze dynamic production processes could not devise a large-scale model compatible with direct digital control. Examples of this failure are taken from three industries: cement, chemicals, and machinery.

Cement Production and Computer Control

From 1963 to 1971, eighteen process-control computers were installed in cement factories in the United States.[1] During this period their performance was at best uneven. The computers in three plants failed to control processes to any advantage. In two instances they were returned to the vendors.

Many of the companies, using X-ray analyzers, developed blending programs, by means of which materials from the quarry were blended in the appropriate amounts before crushing and grinding. Such programs, based on straightforward sampling algorithms, were easy to design, but hard to implement. Sometimes the particles crushed together for the sample were not of the proper size; often the concentration of a particular material was not a simple linear function of the X-ray penetration of the sample wafer.[2]

These problems of implementation were upstaged, however, by theoretical concerns. Factories had to control the temperature, rotation speed, feed rate, and exit gas flow rate of the kiln itself, where the mixture was burned or baked to produce the cement "clinkers," or nodules. Remarkably, although cement production is an industry with a 150-year world history, "the fundamental physical and chemical reactions that occur in the cement process have not been fully understood."[3] Thus, for example, material moving through the kiln is exposed to various heat intensities radiated in complex, little-known ways. Plant operators, who cannot directly measure the application of heat to the materials, may have to choose the temperature at a single point in the kiln as the best indirect measure of effective heat. If their choice is a poor one, they lose some control over the quality of the cement produced. Furthermore, the choice may have to change with the composition of incoming material.

Because of these theoretical difficulties operators could not design effective models of the kiln process and therefore could not use the computer to improve the operation of the kiln. In other words, the computer could not balance such elements as feed rate, fuel rate, and rotation speed according to a model of operation that would maximize kiln output consistent with a specified quality level. All the companies that developed, or bought from outside consultants, such models in the early years had to dispose of them later, operating their kilns by empirical measures (logged by the computer) and a growing feel for the operation of the kiln. Indeed, the most successful companies were those that developed their programs slowly, on a learning-by-doing basis, employing in-house engineers and operators who were familiar with cement and committed to control engineering.

Chemicals

Chemical and petrochemical processes also proved difficult to map.[4] In distillation columns, the relationship between heat input at the bottom and evaporation at the top involved complicated dynamics. Moreover, chemical production presents complexities both in the physics of the process—the interaction of temperature, pressure, and speed of flow—and in the chemistry of the transformation. In the late fifties, for example, engineers discovered the phenomenon of "multiple steady states." In one polyethylene plant the production process functioned within limits for the first five months but then, inexplicably, an explosion took place in the reactor, leaving only charred plastic. Engineers at length determined that the process involved three states, each with its own configuration of important variables such as temperature and pressure. Two of these steady states, however, were only stable in a narrow range of the variables: slight, random variations in temperature or pressure could cause an explosion. Only the excellence of the controls, which had kept deviations within narrow bounds, had allowed the reactor to function as long as it did.

Until World War II these dynamics were not centrally important, since most controls were local, each separately governing only a small part of the process, such as pressure in a column or temperature in the reactor. After the war engineers tried to develop large-scale, dynamic models; but because in a dynamic process few subsystems are independent of the others, it was hard to find a place to start. Some engineers employed a hierarchical model in which the higher system supplied parameter values to the lower ones but did not receive input from them. This method proved unsuccessful: although engineers developed effective models of steady-state behavior in the production process, they could not model the behavior of the system when it deviated from the steady state—when it was turned on, for example. Large central computers could have justified their enormous cost only by providing direct digital control, enabling large models to integrate the separate controls. This is one reason why automation, predicted with much fanfare in the late fifties, proceeded at a slow pace through the sixties.

Machine Tooling

Machine tooling presents similar problems today. Engineers have not yet devised an optimal machining system. In a "totally controlled" machining operation the feed rate of the tool and the turning speed of the workpiece vary continuously with the changing shape of the workpiece.[5]

The fundamental problem here is the wear and tear on the tool. It may seem at first that the machine should run as fast as possible, forcing the workpiece and the tool together at tremendous levels of mechanical force. But then the tool wears out rapidly, and economic gains from fast machining time are overbalanced by the cost of replacing tools. On the other hand, a tool running at some uniform, safe speed during its entire useful life cannot accommodate varying pressures at different points of the workpiece, changes in workpiece materials, or successive phases in the machining of the workpiece. Taylor, obsessed with this problem, spent twenty-six years trying to develop optimal feed and speed rates for machining, finally producing a formula that he was able to convert into a circular slide rule for solution on the shop floor.

It is now apparent, however, that no single formula will lead to optimal machining. Instead, a long-term development program at the shop level is required. As one author writes,

Implementing a completely optimizing system . . . requires experiments to determine wear-versus-time plots for both the tool and the material for at least two different speeds. These are needed because actual values can vary from published values by as much as 20 percent. It is also necessary to determine criteria for tool life and to devise a wear-rate sensor to measure how tool wear changes with such factors as time and geometry of the cut. . . . At present, however, wear-rate sensors do not exist and most manufacturers are unwilling to spend the time to obtain the necessary data for the tool-life equations.[6]

In general, machine tools, "contrary to their advertisements, could not be used to produce parts to tolerance without the repeated manual intervention of the operator in order to make the tool offset adjustments, correct for tool wear and rough castings, and correct programming errors."[7]

The Technical Basis for a Learning Approach

The dead end of large-scale modeling comes at a time when a learning-based system of machine development is possible. Recent critics of technological planning have emphasized this approach in several areas. The authors of a study evaluating current plans for a nationwide automatic aircraft control system emphasize the need for continual modification of such large-scale systems according to the controllers' experience.[8] They suggest that certain complex functions, such as reclassifying planes in the tower's airspace according to new criteria, accommodating instructions to sudden changes in the weather, and responding quickly to the presence of military aircraft, cannot be managed by a computer. Similarly, critics of the installation of flexible machining systems advocate a process of "evolutionary incremental-ism,"[9] in which subsystems are linked by integrating controls only after experience has clarified their behavior.

This evolutionary perspective has a sound technical foundation, for with the proliferation of microprocessors the cost of learning has fallen. As we have already seen, chemical engineers could not justify installing large-scale computers that were unable to integrate the separate controls into a control system. But a microprocessor costing between one and three thousand dollars can be dedicated to a single function or part of the production line. Used initially, perhaps, as a simple analogue control, it will be capable of storing, processing, and transmitting an array of data on operating characteristics and performance features. As operators and engineers continue to analyze and integrate this information, factory personnel can, over time, develop a more thorough understanding of the underlying chemical and physical processes.

In such a context the general issue of integrated control is analyzed into separate problems, each receiving an optimal solution. Then, through a continuing process of modular design, tasks are combined, new solutions are developed, and integrated control systems emerge. As one author observes, "microprocessors are programmed to store their analyses; they provide the knowledge base for their reprogramming, for interactive links, and for much more accurate pinpointing of where additional controls or sensors are needed, or even fundamental research."[10]

This form of automation development may be more effective than installing fully integrated systems. As we have seen such systems still pose difficult quality-control problems. Engineers need a technology that allows them to scale up a shop's investment in automatic machine tools, beginning perhaps with simple jobs that are loosely connected to other parts of the machine-shop operation, then moving on to the more complex and integrated tasks. The minicomputer, the immediate forerunner of the microprocessor, introduced this possibility. The computer console can be placed next to or inside the machine tool and is completely dedicated to the operation of that machine. Such a decentralized control system is likely to shape the firm's investment in flexible machinery systems. Managers investing in machine shop automation can gradually increase their investment as they learn about the characteristics of particular machines and particular jobs. The learning element is given full play as the cost of information acquisition falls.

Integration or Flexibility?

While the technical foundation exists for a learning approach to machine installation and development, there are other barriers to the emergence of an evolutionary framework. The design of cybernetic systems is based on the principles of integration and flexibility: integration, because controls continually regulate the boundaries between parts of the production process, and flexibility, because these same controls create machine systems that respond to changing ambient conditions. But these two principles of design tend to pull in opposite directions. If the engineer integrates all the production parts into one system and designs controls that anticipate all environmental changes, then he comes close to designing the perfect self-regulating machine, the total system that incorporates all relevant forces and processes. He becomes a systems planner and utopian designer. As Boguslaw argues, the engineer invokes an ancient tradition of utopian community design — of philosophers, prophets, and revolutionaries hoping to discover the few invariant principles of human behavior that, if represented in community rules, would create a self-regulating social life, free of conflict and change.[11]

At the same time, the principle of flexibility is exerting an opposing force. Flexible machinery creates a machine system potential, a capacity to produce many different parts, or combinations of parts, and to change the volume of production. Moreover, as the company's market changes, the machine system's distinctive competence will change. Engineers will develop new software, new control programs, and new configurations of the hardware at hand to adapt the machine to its setting. The postindustrial machine evolves. Its design is open-ended.

There is more at stake here than competing philosophies of engineering design. Each principle sets the stage for a different conception of work. The principle of integration and utopian design reinforces a Taylorist view: the more perfect the machine, the simpler and more rational the job. Systems theory, control engineering, utopian thinking, and Taylorist prescriptions all converge to limit the worker's skill. In contrast, the principle of flexibility creates a conception of work in which the worker's capacity to learn, to adapt, and to regulate the evolving controls becomes central to the machine system's developmental potential.

In Sum

Since the beginnings of the scientific management movement, engineers, designers, and managers have tried to codify knowledge in explicit procedures, routines, and machine designs; they have also tried to separate doing and knowing by limiting the knowledge and planning competence of shop floor workers. A critical gap has emerged between the tacit, experiential knowledge that comes from shop floor experience and the theoretical knowledge of machine fundamentals. This division has limited the development of flexible automation systems, for the design and integration of control systems require extremely detailed knowledge of the specific dynamics of a particular machine system, not just of machines in general. As one industrial psychologist notes, "The operator can achieve better results than the engineer. This can probably be put down to his ability, derived from intimate experience of the plant, to take into account the many ill-understood factors which affect the plant's running but which he cannot communicate to the engineer."[12] The tension between learning and control is now evident in the politics of work design.

II

Workers and Machines

7

Beyond Simple Labor

Taylorism in Historical Perspective

The assembly line is the emblem of mechanized industry. The character of assembly-line labor—the pacing and fragmentation of work, the reduced and narrowed role of the worker—typifies the way techno-logical developments have robbed work of interest and the worker of initiative.

I have already briefly described the rise of Taylorism, most concretely realized in the application of time-and-motion studies to the analysis and organization of work. Perhaps no sight is more familiar to the machinist, the textile worker, the auto worker, and the baker than the ubiquitous methods engineer with his stopwatch, observing work on the shop floor, noting the average required speed for each motion and step, and calculating an average rate for each job. Average time not only establishes the criteria for base pay, extra-incentive pay, and job classes, but also contributes to the regimented atmosphere of much factory work. The average time establishes an absolute limit to the worker's control over his bodily movements during the work day.

The struggle between the worker and the efficiency expert is part of the lore of factory life. The methods engineer observes the drill-press machinist. The machinist dutifully places one metal sheet beneath the drill and makes the appropriate neat round hole in the right place. He places the sheet in the bin and positions the next sheet beneath the drill. The methods engineer records the number of holes drilled in three minutes and moves on. The machinist, resuming his natural ways of working, places *two* sheets beneath the drill at the same time.

Indeed, within certain limits a group of workers can control the pace of work in a factory. Few workers want to be "rate busters"; and when engineers time particular methods, most workers will work to the book, carefully, in order to create as low an average time as possible for any particular task. "Organized soldering," as management calls it, remains one of the central features of factory life, a sign of the continuing conflict between worker and manager over the control of daily life inside the factory walls.

In a recent and most influential text, *Labor and Monopoly Capital*,[1] Harry Braverman, a Marxist, has examined the organization of automated work settings in historical perspective. He places Taylorism at the center of work rationalization and the destruction of worker initiative and satisfaction. Although Taylorism remains one of the most important features of factory life in modern industrial countries, one must not exaggerate either its historical significance or its present relevance. Braverman does both.

The efficiency expert did not spring full-grown from Taylor's head, but was rather one element in a larger process of factory rationalization that began in the 1880s and was nearly complete in its essentials by 1920. Factory work in 1880 was organized very differently from that of today. In machine shops, the sources of most of Taylor's ideas and techniques, skilled workers were actually managers.[2] Owners provided materials, tools, power, and the factory building and then subcontracted work to the skilled machinists, who in turn hired semiskilled workers. The skilled machinists, who earned both a daily wage and a profit, completely controlled the production process. In many factories the owners could not even calculate their labor costs, since these costs were determined by the hiring and supervision practices of the skilled machinists.

Modified forms of contracting pervaded other industries as well. Mule spinners in cotton mills hired informal apprentices and paid their wages; foundrymen hired boys or women to assist them in secondary tasks; and potters profited by using helpers to finish pieces.[3] The salaried foreman exercised similar prerogatives: "In many industries he made most of the decisions about how the job was to be done, the tools and often the materials to be used, the timing of operations, the flow of work, and workers' methods and sequence of moves."[4] Remarkably, as late as 1880 industrial capitalism represented in many of its features a vast jobbing system, akin to the old "putting-out" system of rural textile production, in which capitalists organized the flow of money, tools, and material but left planning and execution to an elite of skilled craftsmen who actually controlled the production process.

This system worked with small, local markets and simple, general-purpose tools. But as markets expanded, owners developed uniform pricing and marketing policies and invested greater sums in larger equipment such as stacking cranes, conveyors, and iron-ore scooping machinery. They had to develop greater control over costs to insure a consistently good return on investment. The rationalization of work was thus one aspect of capitalists' general attempts to increase control over the costs and revenues of large-scale industry. Indeed, it was only after 1880 that the specific study of management emerged. Nelson writes, "There was little evidence of any literature on management in the United States before 1870. Only fifteen articles appeared before 1880; after that time the number increased rapidly."[5]

Taylor's work highlights the relationship between rationalization in general and labor-control methods in particular. He developed methods for the measure and design of machining methods—emphasizing the choice of good tools, blades, belting, and layout as much as accuracy of methods—as part of a general plan for increasing the planning functions of management. Taylor believed managers had to develop good systems for inventory control, accounting (particularly important for the allocation of overhead in the pricing of jobs), and the tracing of materials. Time studies had no meaning except as part of a management system within which the evaluation of methods and materials was integrated into the overall control of cost. This required that management dip down into the factory and, through engineers and salaried foremen, take greater control over operations. Skilled craftsmen and foremen-entrepreneurs had to give up their power.

Taylorism spread only slowly through the factory. As late as 1914 Robert Hoxie wrote that "no single shop was found which could be said to represent fully and faithfully the Taylor system as presented in the treatise on shop management."[6] Management resistance was significant here: "Plant managers, particularly those at the lower levels, viewed Taylorism with considerable apprehension and skepticism."[7] Managers were no more ready than workers to change the way they worked. To be sure, the greatest resistance often came from the foremen and the skilled craftsmen. The unskilled workers posed fewer obstacles, however, not because they were innately docile but because they had never been given much leeway under the old internal subcontracting system. Indeed, since successful Taylorism introduced greater certainty and regularity into work and payment methods, some unskilled workers undoubtedly benefited from the shift to more rationalized methods of production.

Taylorism must be viewed in its historical context; it must be seen as part of the growth of large-scale industry based on national markets, long-term investments, and cost accounting. Braverman, ignoring this context, makes Taylorism the outright creation of capitalists, who in their struggle with workers developed time-and-motion studies to rob the workers of initiative and thus of the capacity for revolution. The historical truth is more complex. A developmental process tied up with national growth and integration transformed both manual work and management, eliminating the role of the skilled craftsman-entrepreneur who controlled the production process on the factory floor.

Once we understand Taylorism as a process conditioned by historical developments, we can imagine that it may itself be superseded by other and different approaches to work, not because workers or capitalists will it, but rather because new developments in technology or market structure modify the motives and actions of managers and workers alike. In contrast to Braverman, who sees Taylorism as the sine qua non of capitalist society (though stopwatch methods and task analysis have penetrated socialist economies as well), we can consider it as one particular stage of capitalist development. We can ask the question, In what ways does postindustrial technology modify the relevance of Taylorism to the organization of work?

The Critique of Taylorism

Soon after World War I industrial psychologists began to expose the scientific pretensions of the Taylorist method. Gestalt practitioners demonstrated that one could not divide human motions into their component parts, alter individual parts, and then reintegrate them into an improved sequence of motions. Human motion was an indivisible whole tied to the integrity and flexibility of the human body. Precision of movement based on an analysis of the subunits of human motions could only result in extreme fatigue, high error rates, or both. The decomposition of tasks and skills into their component parts was unnatural. Similarly, rest could not be regarded as independent of activity. Rest and motion were part of a seam of action. The former could not be calculated as simply a subtraction from motion but rather as a necessary "silence" in the orchestration of efficient movement.[8] Moreover, though Taylorist engineers insisted that careful study would yield the best, most efficient way of doing tasks, other industrial psychologists showed that few tasks were done with much uniformity. Individuals, with their own styles of action and perception, find their own efficient ways to perform tasks.

In the most famous series of findings, researchers demonstrated that social factors play a decisive role in determining the level of worker efficiency. May Smith showed how two similar factories can have very different levels of worker adjustment, satisfaction, and productivity, depending on the overall climate and esprit de corps of each.[9] "Repetitive work," she says, "is a thread of a pattern, but it is not the total pattern."[10] In one of her studies telegraphers who planned to stay in their jobs for a long time developed finger cramps, while those who expected job mobility rarely experienced the ailment.[11] Finally, Elton Mayo demonstrated the significance of group experience and social climate with his discovery that young workers in a relay assembly shop at Western Electric steadily increased their productivity as long as work conditions such as lighting and rest periods were being varied. He concluded that the social condition of experimentation itself could raise worker productivity.[12]

The Limits of the Critique

All these criticisms of Taylorism demonstrated that tasks could not be examined without considering their context. There was a wholeness to the work experience, whether represented in the integrity of the human body or in the influence of group experience, that resisted analytic decomposition. Yet by and large such discoveries did not decisively affect the design of work on the factory floor. The gap between theory and practice had many causes. Efficiency experts were engineers of a sort (many came up from the blue-collar ranks), unfamiliar and uncomfortable with the methods of academic psychologists and physiologists. The new findings threatened to complicate their jobs immensely. On the other hand, if the importance of group life was recognized, and if individuals and groups were allowed to discover their own optimal pace and form of work, engineers and efficiency experts would have little to do. Their supervision, time charts, and stopwatches could play only a subsidiary role in the control of production. Supporting the Taylorist work designers in their opposition to the gestalt psychologists were other psychological traditions, rooted in behaviorism and individual testing.

Most important was the attitude of management, which feared it would lose control over wages, profits, and productivity if it gave too much play to worker initiative. There are endless examples of experiments in which groups were given greater control over the pacing of work and individuals were given enlarged jobs that permitted a more varied rhythm of work.[13] But such experiments, numerous and successful as they were, never dominated the industrial climate.

Managers had some reason to fear losing control. William Whyte provides an interesting example.[14] Women paint little toys and place them on hooks that pass by on the way to and through a baking oven. The women complain about the heat and the pacing, and a large number of the hooks enter the oven empty. The foreman installs fans, and the women are happy. The women then ask permission to control pacing. Surprised, the foreman agrees, providing the group leader with a button that allows her to choose among three conveyor speeds. With the women as a group deciding the sequence of speeds to be chosen over the course of the day (high, medium, or low), productivity rises significantly, the average speed of the line increases, and the women are happier. But consequently their bonuses increase, they make more money than some skilled workers in other departments, the latter complain to management, and the system is dropped.

The system works, but it introduces too many perturbations. Beneath the superficial issue of pay inequities are tensions and strains arising from the perceived threat to the entire hierarchical arrangement extending from management through engineers, efficiency experts, and foremen. As control is vested in the workers, the hierarchy loses influence over basic work parameters, and disruptions can develop. Never mind that productivity increases: who is to say that it will stay high, that it will increase even more if necessary, or that workers will not now ask for control over other areas? The implications are too open-ended.

Beyond the Critique of Taylorism

The critique of Taylorism, however correct it might be, had a utopian quality. Wherever large numbers of semiskilled manual workers labored in a framework that limited their initiative and dulled their interest, management always had to employ coercive methods to ensure that the work was done consistently and according to plan. Enter the foreman as "straw boss," pushing the workers, pressing them, fighting against their instinctive group soldering. Any other approach was too uncertain. Remove the straw boss, give the workers initiative, and management could lose basic control over production, control it had wrested from the craftsmen and skilled workers of an earlier time and was not going to relinquish to semiskilled workers now.

The theoretical opposition to Taylorism cannot provide a practical alternative, nor can it provide the basis for a reconception of work in a postindustrial framework. Indeed, postindustrial technology makes both Taylorism and its critique increasingly irrelevant. As we go beyond the old conceptions and realities of industrial work, we are at first unable to imagine a role for the worker in automated industry; yet as we penetrate farther into the problem, we see new possibilities for the mobilization of worker attention and thought that take work design to an uncharted conceptual and practical terrain. Let us explore these propositions in greater detail.

The Relevance of Time and Motion
A semiautomatic steel rolling mill opened in Lorraine, Ohio, in 1949, and nine men were shifted from the old to the new mill.[15] Steel billets were fed to a furnace, heated, discharged, pierced with a plug, and then conveyed to a rolling stand, where the billets were squeezed, stretched, and widened to make steel pipes. A discharge worker controlled the ejection of billets from the furnace with a series of levers and buttons; a piercer sitting in a pulpit operated a set of levers to pierce the hot billets; a mandrel operator watched an automatic machine push a bar or mandrel through the hole in the billet; a rolling-mill operator controlled levers and buttons to squeeze and lengthen the pierced billet or shell; and a stripper watched an automatic machine remove the mandrel. Another worker, the "piercer plugger," prepared the plug for insertion into the billet to create a hole; he was the only one who did heavy manual labor. The rest either operated a series of controls, much as they would drive a car, or simply watched and supervised the operations of automatic systems.

Was the work organized along Taylorist lines and according to Taylorist principles? Motion studies would have no place here. There were no overt physical motions that could be monitored, analyzed, or decomposed. The tasks were chiefly fine perceptual-motor ones, in which the interaction of attention mechanisms, the reactivity and responsiveness of the nervous system, and timing skills were all critical.

Time studies would be similarly irrelevant. Gross production rates were determined by the operation of the machinery, by the continuous flow of hot billets through the different stands or work stations, and by the capacity of the group to coordinate its actions. The discharge operator had to release billets at a rate matching the work of the piercer operator and the piercer plugger. Similarly, when the mandrel operator threw the system into "manual," he had to coordinate his actions with the rolling mill operator. Time studies could not increase the speed of this mill, for no amount of study or coercion could make an operator push a button faster. Timing and coordination, rather than speed and pace, determined worker productivity.

Individual incentives and piece rates would have no place either, nor could any worker be a rate-buster. The steel moved along the conveyor at a rate determined by the interaction between the equipment and the coordination of group actions. Job classifications were similarly problematic. On the one hand, the physical difficulties of the work, and consequently job class and pay rates, actually fell for some of the workers—though group bonus incentives more than compensated for the decrease in base pay rate, since the mill was very productive. On the other hand, the workers felt more engaged and more responsible. Some of them surveyed the whole production process from atop pulpits, watching the flow of steel from beginning to end. In the old mill they had been more specialized in their functions and narrower in their perspectives. Formerly, nine men had worked at the furnace alone, pushing and pulling the billets in and out of the furnace; it was hot, physically difficult work. The old mill's work crews had been organized around separate machines or work stations—the furnace, the rolling mill, and the mandrel mill—and few men had viewed or experienced the mill as an integrated industrial process.

Improving Productivity

Industrial psychologist-engineers entered a rolling mill in Britain in 1962 to study worker productivity.[16] Seventeen percent of the "strip" (the flattened steel) produced here exceeded the desired tolerance. They noted the following: the X-ray gauge that measured steel strip thickness was hard to read and poorly located; the operator who flattened the strip worked with poorly designed dials; the pulpit operators sat squeezed between the pulpit wall and the control desk or console; the pulpit operators could find no convenient place to put the schedule sheets that showed what gauge of steel was being produced; the operators had no easy access to standard tables and thus had to rely too much on memory to determine the appropriate rate of squeeze on the strip for a particular gauge of steel.

The findings and recommendations of the observers are simple and straightforward and make no reference to human relations or worker satisfaction. Yet they do not in any way reproduce or even echo Taylorist design principles. And how could they? Because operators' attention mechanisms and fine perceptual-motor skills depend upon timing and responsiveness, their productivity can be improved only by mobilizing their attention according to the dynamics of their nervous systems and styles of information processing. Dial display, trivial as it may seem, can be critical here, since poorly designed dials in which direction and magnitude are hard to discern can limit the rate and accuracy with which operators observe and process relevant information. A straw-boss foreman who pushes workers to their physical limit would not be effective in this context. While coercion does not disappear—the threat of unemployment always hovers in the background—it plays a less obvious part in the organization of work and the mobilization of effort. New motivations to work must and sometimes do emerge.

What Is Work?

The engineers at the Lorraine rolling mill developed pay rates and job classes on the basis of Taylorist conceptions. True, the measurement of work was a more complex affair. When the piercer was just watching the automatics and observing mill operations in general, was he working or not? The engineer said that when the billet was at his stand he was working; at other times he was not. The workers protested.

We don't think that in a mill like this you can really tell the difference, the way you say you can, between just sitting on your tail in front of a machine—which you call idle time and for which your plan gives no credit—and being alert, thinking, and ready for action—which you call attention time. Take the stripper man for instance. You call it attention time when the pipe is going through his machine and the conveyors around him. You call it idle time after the pipe has left the area. But in those seconds—under some circumstances they are minutes—the good stripper man is on the alert for signals from other areas and planning what to do next. . . . I get my best ideas about the job . . . when I am in the can or at lunch or on the way home.[17]

The engineers replied that the job class reflected this level of attention and thought, but the workers resisted this answer. At stake, as Charles Walker puts it, was not the equity or reasonableness of the particular measurements but "the practice of measurement itself."[18] The workers believed they were in a new situation. Their daily experiences were too vivid to be ignored. Surveying the mill from pulpits, they worked as a group, "tuning" the mill and developing a group rhythm of preparation and action.[19] With fewer workers the foreman is less a straw boss than a technical supervisor, responsible to the machines' and the workers' technical needs. As management relies more and more on the work group for the daily flow of technical information, the distinction between workers and management begins to blur.

Something has changed. The mill operators of Lorraine are no longer manual semiskilled workers. Yet, the engineers ask, are they really managers, supervisors, controllers? After all, they are still tied to the machine rhythm. To be sure, their movements are more subtle, and tension and fatigue are rooted in nervous rather than muscular activity. Like assembly-line operators, however, they still must swing with the machine system. Indeed, while the assembly-line operator could commit his body to the line but free his mind for daydreaming, the rolling-mill operator must commit both body and mind to the rhythms of continuous flow. Workers and managers are engaged in a new conflict, not over the division of prerogatives, rights, duties, classifications, and pay scales, but over the definition of the situation itself. As the conflict moves to a deeper level, both Taylorism and its critique lose their relevance. How can work be understood in this new context?

The semiautomatic rolling mill (semiautomatic because the workers remain at the feedback controls, working them if they must function on manual or taking over when they malfunction) demonstrates that under conditions of postindustrial technological development new definitions of work and new work conflicts will emerge. Studies of other factories and plants confirm this. In power plants,[20] in continuous-process bakeries,[21] in chemical refineries[22] supervision becomes more technical, job classes are reduced in number and reflect more integrated tasks, the number of production workers falls, a distinct work group emerges, and piece rates disappear. As the workers' muscle systems are demobilized, their nervous systems come into play. Deeply involved in supervising the feedback-based control systems, workers face new tensions in the counterpoint of watchfulness and boredom.

Defunctionalization

Viewing the cybernetic machine as a simple, linear extension of the mechanized machine, some analysts fail to perceive its new flexibility; they consider the cybernetic work setting merely an amplified version of assembly-line automation. Thus Braverman argues that automation simply extends and amplifies the "deskilling" process, through which worker initiative, understanding, and conceptualization are progressively inhibited.[23] In this he echoes management analyst James Bright, whose analysis, based on a 1956–58 study of thirteen factories (before any process computers were introduced into American plants), suggests that automation narrows the role of workers, tying them more and more to the machine system and increasingly constraining their freedom of movement.[24]

Both Braverman and Bright have confused two distinct processes. As workers are deskilled, their actions are narrowed and they become more integrated into the machine system. In defunctionalization, on the other hand, the workers give up all execution functions and manual action and are fundamentally displaced from production. Displacement is hidden in semiautomatic systems, as workers lose execution functions but gain in control activity. Though engineers can design mechanical extensions of human arms and legs, they cannot always design extensions of human eyes and brains. There remain technical and economic limits to the design of effective control devices in many factories. Automation leaves gaps. Continuous-process machines take over certain functions, such as furnace feeding and discharging in the rolling mills, but cannot be designed to assume others, such as piercing the billets with a plug. The latter job thus becomes even harder, since the worker can take fewer breaks and must work faster.

These are intermediate developments that obscure the larger question. As all work becomes watching and attending, as feedback-based controls are inserted everywhere, in what sense does work exist any longer? This is not a question of skills, Taylorism, or Braverman's "separation of execution from conception"[25] — there is no execution — but of meaning and identification. What is work? Are machines destined to make work and workers superfluous? If so, we face a question that transcends capitalism and Taylorism and cuts to the core of the organization of social life. Mechanization narrows skills and coerces workers to commit their bodies, if not their minds, to the machine process. Postindustrial technology threatens to throw them out of production, making them into dial watchers without function or purpose. The problem of skills is dwarfed by the larger problem of fundamental identity and function. Again, because Braverman could not see Taylorism as a historically conditioned process, he could not discern the radically different issues raised by the development of automatic-cybernetic, as against mechanized, systems.

The Problem of Machine Failure
If we listen to workers' own accounts of their experiences, we learn of further complexities. Displacement from direct execution does not relieve the workers: instead, they become persistently vigilant and often nervous, feeling that though they execute few tasks they have become more responsible for the entire process. One worker in the Lorraine rolling mill observed, "You have to think more about the job you're doing. You can't look around. In the old mills the job got so you didn't have to think, no mental effort. This job is very touchy, you have to watch all the time, think every minute. . . . Even when the mill is turned on automatic, you still have to think all the time."[26]

Similarly, observers of a continuous-process cookie- and cracker-baking plant found that even though workers grew comfortable with the automatic machines, they still did not feel free to move about. "Any real sense of freedom is illusory, since any failure to observe alertly the various gauges may result almost instantaneously in serious product loss and/or costly machine damage."[27]

The incentives problem in the Lorraine rolling mill is illuminating.[28] Under the group bonus plan the more the steel mill produced, the more the workers were paid. Workers argued that they could not control the rate of production, since so much depended on the machinery. Engineers replied that base pay was calculated to include "average expected breakdown." The workers did not believe this: their experience suggested that automated systems introduced a new level of uncertainty to factory life. Simple machines may break down in expected ways based on design, load, and maintenance practice. Unexpected breakdowns occur, however, in high-speed automated systems, in which a network of electrical, electronic, hydraulic, pneumatic, and mechanical devices is subject to the stresses of vibration, corrosion, and electrical failure.

We see that watchfulness and attention must be mobilized, because cybernetic-automatic systems introduce new and unexpected ways of failing. Work takes on a new meaning in this context. Steady-state functions—the control of expected errors—are turned over to the feedback-based controls; but discontinuities, whether based in failure, in the introduction of new materials, or in the redesign of the machine system itself, draw on worker knowledge, attention, and watchfulness. Thus in cybernetic systems machines and workers complement each other with respect to a typology of errors: machines control expected or "first-order" errors, while workers control unanticipated or "second-order" errors.

If we disregard the developmental functions of workers in cybernetic systems—if we assume, as Bright does, that automation means increasingly self-regulating machinery to the point where discontinuities are also self-managed—then we must anticipate a crisis of displacement and defunctionalization. A choice will have to be made: either workers will remain involved because it is "good" to work, or they will be discarded because they are too expensive and unreliable. This choice will presumably be strictly political: Taylorism and criticisms of Taylorism will have no relevance.

At present, however, the trend toward displacement and defunctionalization is in conflict with a growing mobilization of attention and watchfulness that arises from the imperfections, the discontinuities, of cybernetic technology. Preparation and learning are emerging as core elements of work. If, as I believe to be the case, error is inevitable in automatic systems—if there are always to be modes of failure that cannot be automatically regulated by feedback-based controls—then learning must be instituted in order to prepare workers for intervening in moments of unexpected systemic failure. Failure, in turn, is a specific example of discontinuity and developmental change. Thus we could define postindustrial work as management at the boundaries of systems and physical realities. Historically, we would then see the worker moving from being the controlled element in the production process to operating the controls to controlling the controls.

8

The Failure of Controls: The Case of Three Mile Island

On March 28, 1979, an overheated reactor threatened the health and safety of people living near the Three Mile Island nuclear power plant in Pennsylvania. Long afterward, political debate focused on the pros and cons of nuclear power as a source of energy. But the near catastrophe exposed another, equally fundamental issue. Were the "automatic systems" designed for fail-safe operations and replete with multiple backup systems for "defense in depth" in fact self-regulating? Would the plant prove safe and reliable in the face of component failure, electrical failure, or operator mistakes? Many engineers argued, and continue to argue, that they could design fail-safe systems capable of anticipating and reacting to all relevant contingencies. For a few days engineers and scientists could not explain the sequence of events, however, and opponents of nuclear power, horrified by the accident, argued that no system could be designed that was sufficiently reliable and safe.

Curiously, while both sides debated the merits of purely automatic systems, neither really considered the operator's role in amplifying such a system's capacity to respond to emergencies. Both sides shared the utopian image of the automatic system as a sort of perpetual-motion machine with no need for human intervention. Nuclear-power advocates argued implicitly that such machines were within reach; nuclear-power opponents, that they were not. But neither side dispensed with the image entirely.

The stakes at Three Mile Island were, of course, bigger than at a chemical refining plant, a steam-turbine generating plant, or a rolling mill. Yet the events at the nuclear plant and the responses of the people involved throw light on issues of general importance to those who wish to develop a constructive image of postindustrial work. Chemical plants explode, turbine blades are sheared, boilers crack, steel pipe bends and fractures; lives can be and have been lost in the operation of such factories. Failure is central to automatic systems. By studying the accident at Three Mile Island in some detail we can better grasp the nature of postindustrial work.

The Accident

The design of the power plant at Three Mile Island resembles that of a simple boiler. A reactor core heats up as the radioactive rods within it continuously decay. Water piped through the core absorbs both heat and radioactivity and is not used directly to create the steam that runs the turbine. Instead, the radioactive water remains within a closed system of pipes called the primary coolant system. Nearby, a second set of pipes, which does not run into the core, picks up the heat—but not the radioactivity—from the primary coolant system. Steam from the water in this second, independent system runs the turbine.

The accident begins with a maintenance mistake. Workers are cleaning filter tanks, or polishers, in the secondary coolant system. When sludge sticks in the polishers, they attempt to force it through with high-pressure air, but a faulty valve in one of the polishers allows water to leak into the air pressure line. "The water reaches a junction with another air line that controls the valves to the other filters. There is a sudden change in pressure, and the valves that control the flow of water through the tanks close."[1] The conditions for the accident are now established. The secondary coolant system, now blocked, cannot remove heat generated by the reactor. If the reactor overheats, the core casing may melt, pushing a very large amount of radioactive material through the subsoil and underground water of the surrounding areas.

Engineers have designed the system with backup defenses. First, the reactor core shuts down: rods immediately cover the fissionable material to prevent rapid radioactive decay. Small amounts of radioactivity and heat are still produced, however. Next, auxiliary pumps should automatically feed water to the blocked secondary coolant system. But here human error intervenes.[2] The two valves controlling these emergency feedwater pumps are not operating, although no one realizes this until later. A yellow maintenance tag covers one of the two lights that signal valve closure, and the operator does not see the other light. With no water flowing through, the secondary coolant system cannot accept heat from the primary system. Pressure thus continues to build in the core.

Another backup comes into play. As it is designed to do, a valve on the pressurizer, the auxiliary tank of the primary coolant system, gives way under pressure, venting steam and water into the main drainage tank of the containment. But now another failure occurs: the valve does not close. Moreover, a light on the control panel indicates that the electrical power that opens the valve is off, misleading workers into believing that the valve has finished its work. No one realizes that the primary coolant system is still being emptied.

In one moment the character of the accident has changed. First, increasing temperature and pressure threatened to burst the main pipes, destroy the circulation system, and melt the core. But with the pressurizer valve stuck open, pressure falls as water drains out of the coolant system, leaving the core vulnerable. Even though the control rods are covering the fissionable material, an uncovered core (a core without water circulating through) can overheat to the melting point.

These events trigger the next defense. As the stuck valve remains open, cool water under high pressure is automatically pumped into the primary coolant system. But now problems of diagnosis emerge. A sensor on the pressurizer, which measures pressure in the primary coolant system, shows that pressure there, though it fell during the first seconds of the accident, is now dangerously high. Fearing the main pipe will burst, an operator shuts down one feedwater pump and reduces the flow of water in the other. Without adequate information, however, the operators do not understand that water evaporating from the primary coolant system is entering the pressurizer as steam, creating the appearance of high pressure in the primary system.[3] Thus high pressure to the pressurizer masks falling water pressure in the main pipe. By shutting down one of the emergency pumps, the operators have reduced the flow of cooling water to the core.

Only now do the operators realize that the emergency feedwater pumps have not been functioning. Contrary to their control panel indicators, water has not been moving through the secondary pipe to remove heat from the main pipe. This revelation has two effects. First, the operators lose confidence in the controls and in their own responses, which were based on incorrect information.[4] Second, they renew their efforts to lower the pressure in the main coolant system. Though the problem has already changed from one of pressure buildup in the pipes (when the feedwater pumps did not come on at the very beginning) to depressurization (as water and steam spill from the main pipe), the operators have not shifted perspective accordingly. They are stuck in a conceptual framework that quickly becomes obsolete as the accident progresses.

This sequence takes place in less than ten minutes.[5] Nearly an hour later the buildup of steam in the main pipe creates bubbles that cause the main cooling pumps to vibrate dangerously. Still unaware that the reactor core is loosing water, the operators shut down the pumps in order to save them. The temperature in the main pipe increases. The intense heat splits some of the water into hydrogen and oxygen, creating a large hydrogen bubble that threatens to block the flow of cooling water through the pipes.[6] Over the next few days a meltdown is feared. The public drama of Three Mile Island begins.

Failure Modes: Materials, Load, and the Human Community

The event at Three Mile Island reveals the many independent and interacting ways in which complex systems can fail. The accident reached a turning point when the relief valve on the pressurizer remained stuck in the open position. As we have seen, this turned the accident from a problem of pressure buildup to one of depressurization. Valve failure is common enough. The valve at Three Mile Island failed because a broken electrical connection kept it open. In another reactor a similar valve failed when a buildup of boron crystals (used to control the radioactive decay process) bent its springs.[7] Valve failure is simply one example of the subjection of materials to use, stress, metal fatigue, load variations, and unexpected environmental pollutants.

Control engineers, who tend to think in functional terms, often forget that the most complex of designs are realized in humble joints, welds, pipes, seams, and circuits, which may decay and fail. Steam-turbine failures have been traced to subtle degradation of materials. In one accident in which two men were killed, nine were injured, and a turbine and generators were destroyed, small black particles of iron oxide had unexpectedly formed in the oil used in the pumping system. Investigation revealed that while the system was shut down for maintenance, a minuscule amount of saline solution had been accidentally and mysteriously introduced into the oil pipes. When the pipes were drained of oil, the solution, exposed to the air, formed black iron oxide. Carried in the oil, the oxide spread throughout the system as soon as the turbine was started up. These particles, accumulating in the spaces between pistons and cylinders, were rubbed onto the metal surfaces as black deposits. This gradual buildup blocked the operation of the governing pistons that controlled the flow of steam to the turbine. The investigators concluded that under these conditions the governing valve could not close. Consequently too much steam reached the turbine and generator rotors, which therefore rotated too quickly and broke apart.[8]

Although this situation may appear unusual, it illustrates the impossibility of fully predicting and anticipating the operation of material systems. An organic analogy is useful here. The human body, which regulates and repairs itself better than any machine, is still very vulnerable to small material disturbances (viruses, for instance) that take advantage of the interconnected pathways of the body and ramify throughout the system. Machines too are subject to material disturbances. Some are induced environmentally. A pollutant may contaminate the water that runs through a boiler system; electrical currents between sea water and aluminum brass heat exchangers on ships may induce chemical reactions that crack the tubes. In the steam-turbine accident described, materials accidentally introduced reacted chemically with machine components and fluids.

At the Indian Point reactor in New York salty river water, used to cool the air around the reactor vessel, leaked into the containment.[9] This posed unanticipated dangers to the vessel, since salt can propagate cracks in metal. In another reactor the chemical composition of coolant water rusted the steam generator, causing leakage from the primary radioactive pipes to the secondary clean water pipes.[10] Finally, engineers are discovering that radiation bombardment in reactor vessels has made their steel casings unexpectedly brittle.[11] This raises new problems in the management of reactors. Workers trying to cool an overheated reactor may inadvertently exceed the breaking point, cracking the vessel itself. The half-life of nuclear reactors may be much shorter than anticipated.

Other disturbances arise from the mechanical structure of the machine system. An uneven flow of steam can set up vibrations in turbine blades that may cause them to snap.[12] Similarly, power cables will vibrate and may snap in very cold weather.[13] Gradual chemical corrosion induced by foreign materials may stress metals to the breaking point. Finally, in the Three Mile Island accident the formation of the hydrogen bubble threatened to block water circulation. Engineers had not anticipated a sequence of operator mistakes, combined with several physical malfunctions, that would together create intense and prolonged heat in the core. Consequently, when troubleshooters arrived at the plant after the accident, they lacked the tools, measuring instruments, and techniques for reducing the size of the bubble.

In general, structural engineers still have no clear understanding of metal fatigue and cracking. Tests are often insufficient. For example, despite extensive testing, Havilland Comets (aircraft designed shortly after World War II) exploded due to metal fatigue.[14] As one author writes, "metallurgists and engineers attempt to predict, from limited past experience, the behavior of materials in uncertain future environments of use and abuse. The slightest deviation from experience in a new design can have unanticipated consequences."[15]

Any machine with interconnected hydraulic, pneumatic, electrical, and mechanical subsystems that is exposed to a range of chemical and mechanical stresses can and will fail, often in very unexpected ways. The systems engineer focuses on only one particular version of the machine system—a machine designed to execute certain functions, protected by certain backup defenses. But the engineer cannot imagine all the other systems that intersect, overlap, and are integrated with the designed system. Among these are environmental conditions, unexpected variations due to outside disturbances (such as a fluctuating demand for electrical power), and the peculiar interactions that take place between subcomponent chemicals.

The organic analogy is again appropriate. A complex variety of interconnected bodily systems (chemical, electrical, and muscular) that are exposed to continuous use, stress, and load variation will break down (become diseased) in unexpected ways. Moreover, such breaks may be preceded by subtle, unrecognized indications of stress. In the turbine example, the operators had noticed some sluggishness in the governor's control of the valves, yet they had no framework within which to interpret that sluggishness. Similarly, maintenance workers had observed the black particle buildup in some of the cylinders, but because the particle deposits were so hard, they mistook them for indications of a burnished cast-iron surface.[16] Operators, maintenance workers, and supervisors could not recognize these signals of incipient malfunction, much less determine their causes. Even when the signs of failure are emerging, little might be done precisely because this form of failure is unexpected.

The second failure at Three Mile Island was based on maintenance errors in particular as well as general operator interaction with the machinery prior to the accident. While it may be convenient to blame maintenance workers, both for forgetting to close one air valve and for neglecting to reset the feedwater valves, no amount of discipline or motivation will eliminate such errors. Manuals, protocols, and job instructions may dictate a range of procedures, but in the real world, errors of omission and commission may be caused by job turnover, the stress of shift work, backbiting among workers, the pressure to complete a maintenance job and return the system to service, a supervisor out sick, disgruntled maintenance workers, an operator who is ill or asleep at the controls—in short, all the contingencies and events of human living.

We often find the tendency toward error exacerbated by the social organization of jobs and skills and the information relationships that shape cooperation at work. In a unionized nuclear plant an engineer who wishes to recalibrate an instrument must call a mechanic to turn the adjusting screw. If relations between the engineers and the workers on the shift are bad, it may take some convincing before the mechanic complies. If the instrument is indeed faulty, the reactor may enter a dangerous state while the two bicker.[17]

For example, an aircraft controller in Miami noticed that a plane was losing altitude. In the status hierarchy of the air service system, pilots have more prestige than controllers do. Consequently, controllers hesitate to remind pilots of obvious facts or errors. The controller radioed the pilot: "How are things coming along out there?" The pilot, preoccupied with repairing a circuit and oblivious to his loss of altitude, answered: "Okay, we'd like to turn around and come back in." The plane crashed.[18]

Similarly, in the 1982 National Airport crash the copilot did not aggressively challenge the pilot to reconsider his decision to take off despite serious misgivings—there was ice on the wings. Instead he posed his grave doubts as tentative questions ("that doesn't seem right, does it?"), allowing the pilot to continue his own self-deception that conditions warranted takeoff. The social relationship between pilot and copilot led to a tragedy.[19]

Sometimes actions that contribute to failure appear to be innocent and inconsequential. In one study military maintenance personnel, desiring access to a computer, kept the door of a panel open during two hours of maintenance practice. As a result changes in the air flow through the machine damaged some computer "flip-flop" cards.[20] Later, when some of the cards were examined to discover the sources of computer malfunction, several were found to cause trouble only in that particular computer. The authors of the study further note that when the front cover of a computer console is "properly in place, the location of the power protection panel is such that if the console operator crosses his legs, he is very likely to damage or turn off the circuit breaker on that panel."[21]

The organization and execution of maintenance work presents special problems in the relationship between design and use. Thus, for example, investigation revealed that the June 1979 crash of a DC-10 airplane in Chicago, caused by failed hydraulic systems, was ultimately due to the maintenance workers' practice of removing the engine and the pylons together; in doing so they had fractured a pylon.[22] Similarly, investigators analyzing a succession of military plane crashes found that maintenance workers were reversing the trim tabs that controlled the flaps. The pilot-controlled wires sent the flaps up when they were supposed to go down. Nothing in the design of the plane suggested to workers that they were making a mistake, however; they would have had to be intimately familiar with the design to have avoided reversing the tabs.[23] Finally, one study of a mandatory inspection program suggested that the number of accidents due to brake failure rose after the inspection program was instituted. Mechanics were breaking or weakening parts of the brake while inspecting it.[24]

The foregoing examples demonstrate the difficulty of anticipating the many ways in which particular systems will be used and manipulated. This is a familiar problem in all design practice and policy. A city sympathetic to the disabled creates slopes in the sidewalk where it meets the roadway at intersections so that people in wheelchairs can cross the street with ease. Later, planners discover that blind people, who use their canes to discover where the sidewalk ends and the roadway begins, are fooled by the slopes into stepping into oncoming traffic.

In an urban cleanup campaign trash cans are placed all along ghetto streets. But families crowded into small apartments discard more trash per household than in other areas. At the same time garbage collection in the ghetto is less frequent and less efficient. People who formerly let their trash build up in their houses now place the trash in the street cans, which quickly overflow. The streets are dirtier than before.

The problem lies in the interconnectedness of systems. Error must happen when human complexity confronts the technical complexity of the machine system. No designer can anticipate those novel uses, behaviors, and situations that are not precisely errors, but rather are intersections of human systems with technical systems that create unexpectedly dangerous situations. The real world with all its complexity constantly intrudes. The machine system does not function in isolation according to its own autonomous laws and rhythms; as an extension of the human community it is vulnerable to human failures, flexibility, and imagination. For example, industrial engineers discovered that an automatic stocking maker, whose operators pulled levers in a completely predetermined sequence, produced different stocking weights depending on which worker was operating the lever. The variability of human behavior was somehow transmitted through the automatic machine.[25] The connections between human and machine systems run deep. Indeed, as automatic systems are developed to control the most obvious kinds of errors, the patterns of social organization become major determinants of success or failure.

Flexibility

There is every indication that machine systems will become even more vulnerable to failure in the future. Two factors stand out. First, a flexible system with a wide range of functions, whether in producing chemical compounds, machining engine blocks, or transmitting power, is flexible precisely because there are many different routes or pathways through the system. This versatility, however, may turn against the machine, because the effects of errors, blockages, and breakdowns will spread through the system in many unexpected ways. In the Ginna nuclear reactor in Rochester, New York, a tube holding irradiated water burst, spilling its contents into the clean water used to power the generator. Sensing a pressure drop in the irradiated primary cooling system, the automatics pumped water into it. But because the tube was broken, the rise in water pressure pushed irradiated steam through an open valve into the clean water system and into the air outside the containment. The Nuclear Regulatory Commission suggested that a new type of accident based on the punctured tube had emerged. "If the leak were not stopped or additional cooling water were not made available, then eventually all of the HPI water [water pumped under high pressure to the reactor] would be exhausted and a net loss of primary coolant would occur. Without corrective action, core uncovering would occur."[26]

Second, as we have seen, zones of unexpected failure can be created by external stress or variation, such as startups, shutdowns, variations in the load or demand on the system (for example, how much electricity a power plant is producing), continuing changes in design, and the frequent replacement of worn-out parts due to high production runs. Automated systems, which now rarely fail because of simple mechanical or electrical flaws, are used intensively and are exposed to environmental stresses. These stresses in turn create new sources of uncertainty and new failure modes.

We can find an analogy in daily life. A young child, learning to walk, constantly trips over her own feet. Once she has mastered walking, she may still hurt herself; indeed, because she has mastered walking, she enters new environments that strain her skill in new ways. Each increase in self-regulating capacity is matched by a new context that stretches the newly developed capacity to new limits. Thus the system, always functioning at its limits, is always vulnerable to failure. The failures themselves will no longer be of the simple kind attributable to single causes, but will be traceable "to a number of contributing factors. . . . In many cases only an improbable combination of events (improbable from the point of view of the designers) is consistent with known facts."[27]

The relationship between context and failure plays a decisive role in the obsolescence of software. Studies show that the number of bugs to be found in a given program declines at first but later begins to rise. As old and new users become more sophisticated, they begin to experiment with innovative uses of the program and in the process discover more structural errors.[28] Consequently, the cost of monitoring the program begins to rise. The total lifetime cost of maintaining a program will exceed the cost of developing it by 40 percent; at present some companies are spending more than half their budgets just to keep existing programs running.[29] When the underlying logical design of a program can no longer satisfy the users' demands, the program has become obsolete. It is as if an office space were used for so varied a set of functions (private office work, meetings, parties, renting to outsiders) that at some point the entire space would have to be redesigned to accommodate the new and richer mix of uses.

Instrumentation

Operators live in a world of signals and signs rather than objects and materials. They observe production processes through information channels that translate unseen events into data. But signals must be interpreted: they only represent the production process. Indeed, workers do not observe separate controls as separate bits of information to be integrated into a theory of what is going on. Rather, they bring a pattern of awareness to the task so that looking and integrating are simultaneous.[30] Thus the interpretation and the observation of signals codetermine one another.

At Three Mile Island the workers did not realize that the pressurizer valve was leaking hot water and steam. In the containment, however, a temperature gauge at the juncture where the pipe from the valve entered the drain tank read above 200°F, indicating a leak of very hot water from the primary coolant system. Yet operators misinterpreted the signal, reasoning that the high temperature was a sign of residual heat due to a single opening of the valve. Others argued that the gauge had read high in the past because of normal slow leaks from

the valve.[31] No gauge can signify anything by itself; it must be interpreted in a logical context.

A similar misinterpretation occurred at the Indian Point reactor in
New York. A worker inspecting air cooling pipes around the reactor
vessel jiggled a float that measured possible water leakage into the
containment. An indicator light on the control panel lit up. Over the
next two weeks, whenever water leaked into the containment building
from a corroded fitting in the fan cooler, the indicator light would go
on again. Yet workers assumed that the earlier jiggle still accounted
for the warning light on the panel. (They were also relying on two
automatic sump pumps in the containment to push out water, not
realizing that both pumps were broken.)[32]

The failure of the attempted helicopter rescue of Americans held
hostage in Teheran in 1980 may well have been the result of similar
problems of instrumentation. The helicopter control system is designed
to signal the pilot automatically in case of blade failure. In flight,
however, sharp changes in temperature can trigger the alarm. There
is some speculation that the pilots in the Iranian desert responded to
a false alarm and grounded the helicopters.[33]

The problem of interpreting signals reflects inadequate linkage of
the social and technical systems of the plant. People can only work
effectively if they work in a climate of trust. They must believe that
the information they are getting is accurate and consistent and that
it reflects the shared values of the work group. Control panels alone
cannot provide this assurance. Thus, for example, a light went on to
signal a quality-control problem in the assembly area of a Volvo plant.
Workers refused to respond to the light; they wanted the foreman to
commit his reputation for competence to the claim that quality problems were emerging.[34] Work designers must find ways to facilitate
the acquisition of trust in symbolically mediated work settings.

The Problem of Models

Finally, failures occur because designers cannot fully anticipate the
many different ways in which machine systems may operate. The
logical structure of the control system may harbor a hidden contradiction of design and function. During the Three Mile Island emergency
the containment was isolated to prevent irradiated water from leaking
into an auxiliary building that could vent radioactive steam into the
atmosphere. But the mechanism for automatically isolating the containment was designed to be triggered by pressure readings, and
workers had reduced the pressure in the containment by turning on

the cooling fans.[35] At Indian Point there were instruments designed to signal water buildup in the containment by measuring humidity. Because the automatic dehumidifier was working continuously, they could not signal that water was leaking from the fan cooler.[36] Similarly, at the Ginna nuclear reactor operators could easily have used a valve on the pressurizer to equalize pressure between the primary and secondary systems, thereby stopping a leak from the former to the latter. The automatics, however, programmed to close the "nonessentials" in an emergency, shut the air valve controlling the pressurizer valve.[37]

The broad features of the Three Mile Island accident were never anticipated. Prior contingency analysis assumed that an accidental loss of coolant from the core would be sudden.[38] Instead the loss was gradual, creating a degraded core rather than a meltdown. Entirely new contingencies appeared. In particular, operators faced the unexpected problem of managing a hydrogen bubble that threatened to block the circulation of coolant water. In the first few days of the accident some scientists feared that the bubble might explode. The fear proved groundless. This anxiety indicated the confusion of scientists and engineers.

Similar problems have emerged in the development of maintenance protocols. Designers try to simplify and rationalize maintenance by creating clear test routines for circuitry as well as for disposable parts. Often, however, attempts to predict the patterns of maintenance work fail. Engineers at a switching center near Chicago discovered that "nearly 40 percent of the possible troubles were simply not findable by the test routines, even though the prime equipment involved large arrays of digital modules [independent subunits] and a sophisticated test program."[39] Even when the problem is a simple one requiring only the replacement of a faulty equipment module, "subtle time interactions lead to equipment states that are not easily interpretable by the test routine."[40] Moreover, disposable parts may prove too expensive. In addition, design changes can make old protocols and test routines obsolete. Thus maintenance engineers for the Apollo program discovered that no manual could adequately instruct workers in the repair of various malfunctions.[41]

Finally, designers away from the shop floor often lack a working knowledge of the products and processes they are designing. In one British machine shop a programmer created the instructions for machining the iron igniter of a gas turbine. He misplaced a decimal point, and the igniter was made ten times too big. When the machinist brought the igniter to the programmer, the latter commented, "That's a very fine igniter we've made."[42]

Hubris and Utopianism

In their search for fail-safe systems engineers demonstrate the hubris of most design professions. The designers of a machine, a building, or a policy are attempting to imprint their minds on other people's lives. Behind the hubris of design lies a deeper utopianism, however, and that is why the designers of machines and the critics of machines share the same world view. The engineer advocating a defense-in-depth system for nuclear plants and the ecologist questioning nuclear plant safety both invoke the ideal of a riskless world, a perpetual-motion machine—whether in the form of perfect technology or of undisturbed nature—where human intervention has no place. Theirs is a deep and primitive impulse that reflects the continuing immaturity of human societies, a wish to make sense of the sometimes benign, sometimes harsh, but always impressive natural environment. Engineers may appeal to machine designs, but machines, as extensions of human communities, are as vulnerable as ourselves. Ecologists may appeal to nature, but nature promises nothing: floods, earthquakes, global climatic changes, potential cosmic catastrophes highlight our struggle against nature as well as our union with her. The wistful utopianism of both the machine critics and the machine defenders reflects a shared desire to escape from the contingency and reality of evolving human communities. Machine systems inevitably fail, given the realities of materials and human behavior. Once we accept failure as a part of technological reality, we will gain a clearer perspective on postindustrial work.

9

Technology, Consciousness, and Developmental Work

The accident at Three Mile Island took a decisive turn when water and steam gushed through the open valve in the main pipe, pressure dropped, and the amount of water flowing through the core to carry away residual heat was dangerously reduced. Initially the reactor was threatened with a pressure buildup as feed water to the secondary pipe was cut off and water in the main pipe became dangerously hot. But the situation changed when the relief valve in the main pipe opened, as it should have, but then did not close. Yet reports and analyses suggest that the operators did not make the appropriate conceptual shift. They remained fixed on the problem of high pressure and consequently took actions that solved the wrong problem and thus increased the chances of a core meltdown.

It is possible, of course, to blame the designers of the plant as well as the workers. With no direct indicators available, workers had to infer the water pressure in the core from the pressure in the core's auxiliary tank, the pressurizer.[1] The designers should also have provided more immediate and direct measures of the status of the main pipe's relief valve.[2] Operators did not isolate the valve and block the drain until two and a half hours into the accident. There were other ways to infer valve failure, however: operators could either read a temperature indicator on a discharge pipe through which water drained or note the water levels and pressures in a quench tank below the discharge pipe. Technical literature published before the accident suggested that pressure in the auxiliary tank might not indicate pressure in the reactor core because of a range of particular dynamic conditions, such as water flashing and degassing of liquid.[3]

In truth, no accident has a single cause, and no particular subsystem is entirely to blame.[4] Display and design limitations, which reflect the impossibility of anticipating all failure modes, and operator error were all responsible at Three Mile Island. Operators could have discovered the open valve problem early in the accident by reading subsidiary indicators, but because they had no clear conception or theory of the accident, they failed to examine these indirect measures. Designers try to provide those displays that seem most relevant, anticipating the major sources of failure as well as the operators' likely information needs. Theory thus guides the choice of information provided. In an unexpected situation the operators must undertake a more open-ended search, and the designer must hope that the information provides a data base rich enough for the development of new hypotheses.

Preconception and Habit

Airplane pilots learn to ignore automatic warnings if they are busy with other tasks or simply habituated to the signals. In the Miami airplane crash described in chapter 8 the pilot, preoccupied with fixing some cockpit circuitry, failed to notice an altitude warning chime signaling that the plane was falling.[5] At Three Mile Island it was only after the first ten critical minutes of the accident that operators discovered that a yellow maintenance tag on the control panel covered the light signifying closure of the feedwater pump valves.[6] They had no a priori reason to look for the tag.

It seems that when the amount of information available exceeds one's processing capacity, one's attention will select only those signals previously deemed relevant to the task. The focus on design limitations at Three Mile Island is reasonable enough: engineers can use such information to improve future designs. But we risk falling into utopian thinking if we believe that good design can eliminate all contingencies and all nonroutine operator decision making. Once again we are forced to consider the contingent, the improbable, and the unanticipated.

The operators at Three Mile Island fixed on their first theory, pressure buildup, despite accumulating evidence that the structure and direction of the accident had changed. The power of these first impressions persisted well into the first day of the accident. James Floyd, the control room supervisor, who entered the control room only after the first hours of the accident, later told the president's commission investigating the matter that he had known before he entered the room that the core had been damaged and that he had presumed all the operators knew this as well.[7] Indeed, he discovered only at the hearings themselves that the other operators had not known the core had been damaged. This suggests that at least an hour into the accident, after the feedwater pumps had been turned back on, the operators did not understand the roots and implications of the initial drop in pressure and the consequent temperature rise in the core. Had they understood, they might not have cut off the emergency water supply to the core—an action that damaged the core further and contributed significantly to the severity of the accident.

The operators were poor diagnosticians. Relying too much on first impressions, the victims of their own tunnel vision, they focused too much attention on the pressurizer: they maintained their conception of the event even when other available information suggested a different analysis. What, then, constitutes diagnostic skill? What features of thinking and acting allow people to respond to uncertain, contingent, and stressful situations?

It is clear from the findings of the president's commission that workers were limited by habit in their response to control problems. Like an instinct, a habit unconnected to a wider frame of reference becomes embedded in the nervous system, untouched by changing circumstances. The operators' habitual mode of response was evident in the one-step reasoning they used to cope with the emergency. Two instances were critical. Operators assumed that high pressure in the pressurizer meant too much water in the system, despite evidence to the contrary.[8] This reasoning limited their ability to imagine a more complex set of events that might produce the same outcome. Similarly, when the pumps vibrated, operators followed the same one-step logic and shut them off. They could not see that the pumps were vibrating because they had not been turned on.[9]

Operator training at Three Mile Island reinforced this conservatism. "During exercises that involve abnormal operations, both during training and examinations, emergency systems are actuated and perform as expected every time. Instrumentation also responds in the correct manner during the course of an abnormal or emergency operation. However, abnormal operation of emergency equipment during normal and abnormal events is not stressed."[10]

In one study of workers response to operating problems in nuclear reactors, the operators' fundamental conservatism led them only to ensure that major variables were in bounds, without further diagnosis. Often the operator responded to a specific symptom without waiting for complete information about what had occurred.[11] Such behavior may be effective under familiar, predictable conditions; it can be dysfunctional under novel ones. At the Ginna nuclear reactor in Rochester, New York, workers left the emergency pumps on in order to maintain pressure in the primary coolant system. But irradiated water flowed through a broken tube into the secondary coolant system. Increased pressure in the latter opened a valve, releasing radioactive steam into the atmosphere.[12] The workers' excessive conservatism, their commitment to guarding the "fundamentals" in a narrow context, created more problems. What kind of training can help operators resist the pull of habit?

Synthetic versus Analytic Reasoning

Workers' commitment to habit may result from analytic, as opposed to synthetic, reasoning. Textbook knowledge of production systems is taught analytically: the operator learns that if a certain pipe is fractured, the water pressure will drop, creating steam in the pipe that may block circulation. The operator learns to reason from underlying causes (fractured pipe) to overt symptoms (blocked circulation). In operating a plant, however, the worker must reason from symptom to any of several possible causes. Tracing the effects of an underlying cause involves analytic reasoning, which can be taught in textbook fashion, much as one would teach the steps of a geometry proof. But diagnosing symptoms requires synthetic reasoning, in which the range of possible causes is narrowed as more information is received. This process is not linear but cyclical, and it requires the worker to move beyond one-step logic.

Synthetic reasoning is more difficult than analytic reasoning. There is no abstract thread of connections to follow; instead, the logical strand is continually interwoven with real information. In analytic reasoning one simply traces a given problem through the appropriate connections; in synthetic reasoning one must determine what kind of problem one faces and which information is relevant. We see that this diagnostic skill depends on an ability to frame problems, infer causes from symptoms, and check resulting hypotheses against one's analytic knowledge. What is the basis for this kind of skill?

We might propose the craftsman as the ideal skilled worker; but highly specialized skills can be limiting. Craftsmen must attend so closely to the changes in immediate material conditions that they cannot survey the entire process from a more comprehensive perspective. For example, iron puddlers, who melted iron to rid it of impurities, had to carefully monitor the flames breaking through the slag, regulate the vents of the furnace to get the best draft, stir the puddle of iron so that spikes of pure iron would not burn upon contact with the gases, insure that pasty masses of iron at the bottom of the puddle did not chill and stick, form the pure iron balls ("blooms") of equal weight, and squeeze the slag out of the balls.[13] The work took concentration, attention to detail, and responsiveness to the changing conditions of each batch of pig iron. The method of puddling was fixed in its broad features, but each puddling task required careful attention and planning. The puddler exercised judgment, but within a very narrow domain, and was chained to the physical and chemical processes he so closely monitored. Now the process of continual adjustment is controlled by technology. The operator, freed from this close level of scrutiny, can lift his head above the details of execution and diagnose the broader features of the production process.

Some analysts urge that operators be taught more theory, that they understand the physics and chemistry of the processes they manage. But, like craftsmanship, theoretical understanding alone is insufficient. At Three Mile Island the lack of theoretical understanding was not an impediment to coping with the threatened loss of coolant water. While the management of nuclear reactors is affected by the complex dynamics of radiation, operators must normally deal with the more easily understood concepts that link the pressure, temperature, and flows and levels of liquids, since the reactor is basically a large boiler.

In the case of chemical plants, studies have shown that operators need not be taught chemical and physical theory or given abstract presentations of plant dynamics.[14] More useful are rules of thumb that guide them from symptoms to causes. If operators use the rules consistently and with success, they develop an analogic understanding of the plant without necessarily knowing why the rules make sense. This allows them to expand the breadth of their control without having to increase the depth of their theoretical understanding.

Skill: A Unitary Definition

Diagnostic skill is based on integration. The density of perception that the craftsman brings to the process—his feel for it—is combined with the breadth of operation that a heuristic approach brings and the theoretical depth that analytic training brings. The splitting of these features of a unitary skill limits conceptual flexibility by severing the link between fringe awareness and selective attention.

Think of the expert driver of a standard-shift car. Her short cycles of action—shifting the gears, placing her foot on the clutch, adjusting the wheel to changes in road conditions—are all made at the fringe of consciousness. If the wheel seems loose, the clutch sticks, or the gearshift seems tighter, she will notice. Similarly, if a tire is losing air or the road embankment is flat, she will feel these problems through her instruments—the wheel or the brake. Displacing this information to the fringe of awareness, she places the driving task at the center of consciousness. Without losing sight of her long-term goal, to go from point to point while observing all the driving rules, she is prepared to respond to the unexpected. Under novel conditions, continuous feedback at the fringe of awareness and conscious planning at the center of awareness enable the driver to respond flexibly.

Two recent studies affirm the significance of fringe awareness. The first, involving air traffic controllers, concludes that "holistic knowing, which comes from an intimate involvement in every detail of the [air] traffic control process, may be necessary to sustain complex control behavior."[15] This intimate involvement must not, of course, led controllers to neglect their primary task, planning takeoffs and landings. Rather, controllers can be intimately involved and plan ahead if they can displace that involvement to the periphery of their attention, while at the same time monitoring all the activity in this fringe area of perception. The second study recommends, on the basis of experiments with airplane pilots, that operators of nuclear reactors demonstrate "blunder-free residual attention before or during their training."[16]

Diagnostic skill is based on just such an orchestration of attention. We rarely solve problems by systematically analyzing all the alternatives, testing them, and then choosing the best one. This mechanistic model bears little resemblance to human problem solving. Instead, our selective attention is based in serial thinking and conscious verbal reasoning. Our fringe awareness, barely accessible to verbal reasoning, brings us gestalt patterns that suggest possible contexts for problem definitions and solutions.[17] This orchestration of consciousness (related, perhaps, to the lateralization of the brain) allows us to engage simultaneously in gross decisions about wholes and contexts and fine decisions about cause, effect, and specific system relationships.

Fringe awareness, attuned to signals of anomalous events, produces insight into new patterns. Imagine entering a room in which an extended family is meeting for a holiday celebration. Although there is latent conflict in the room, because it is a holiday you are disposed to see interactions and communications between family members as signs that people are having fun. Your disposition to view the setting as a happy one is weakened, not when you decide to test your theory of the event but rather when small bits of disqualifying evidence enter your consciousness. Thus you may notice that when a grandchild spills his soda, tension suddenly spikes in the room, and people shift about so as to create an empty space between his grandmother and his mother. Someone makes a joke, the tension falls, and a cheery veneer covers the accident. If you are disposed to respond to novelty, you begin to interpret this as a critical incident—an event with a latent structure that suggests a different impression of the family gathering. Applying this new gestalt, you might reconstruct the sequence of cause and effect to accommodate the critical incident. Perhaps the child's accident, the supposed result of the mother's negligence, is an occasion for the grandmother to express anger at her daughter for having left her husband.

Fringe awareness and selective attention are integrated when the operator can integrate the three modes of knowing: dense perception of physical processes, heuristic knowledge of production relationships, and theoretical understanding of the production process. Heuristic knowledge helps the operator make normal production decisions while paying conscious and selective attention to long-term goals, such as quality and timeliness. Density of perception supplies fringe awareness with anomalous data—data that might otherwise go unnoticed because it is unexpected. Finally, theoretical knowledge helps the operator understand the anomalous data so that he can overcome previously established rules of action and create new ones appropriate to the novel situation.

Failure Modes

Learning stops when selective attention and fringe awareness are in conflict. When selectivity and flexibility do not enhance one another, selectivity becomes rigidity and flexibility becomes chaos. This breakdown was evident at Three Mile Island. Operators, dazed by too many simultaneous alarms, found the control panel useless. Its poor design and incoherent messages mirrored the operators' own confusion. One coping strategy was simply to deny the evidence of danger. When hydrogen gas exploded in the containment building, operators assumed that either a ventilating damp had slammed or the pressure-measuring instrument that spiked with the explosion had simply malfunctioned.[18]

To diagnose problems correctly, the operator's attention must incorporate density of perception, breadth of supervision and control, and depth of theoretical understanding. The lack of any one of these factors produces particular patterns of failure. When all three are underdeveloped, as was the case at Three Mile Island, habit dominates operator responses and the work team is least flexible. When neither a fringe consciousness nor an integrating selective attention is developed, the worker typically feels bored and valueless under normal operating conditions and overloaded under stressful conditions; indeed, the former can lead to the latter. Training at Three Mile Island, limited to some introductory theory and "single-fault" response requirements, reinforced habit. While pilots, trained on expensive simulators that mimic the operations of a real plane, are taught to cope with a wide range of failures, many nuclear plant operators, after a brief course of formal instruction, are tested by the Nuclear Regulatory Commission by "walking through" a response to an emergency.

When density of perception is not combined with breadth of operation and depth of understanding, then the operator is overwhelmed by detail. Events and perceptions that should be at the fringe of consciousness move toward the center, limiting the worker's ability to attend selectively to a succession of new details that may confirm or invalidate his initial impression. Breadth of operating responsibility without theoretical depth or perceptual density may limit both the range of fringe awareness and the power of selective consciousness to formulate new hypotheses. An operator depending on heuristics may be unable to formulate a conception of an emerging problem.

The limits of heuristics become most apparent when the automatic controls fail infrequently. One analyst has observed: "The situation will be characterized by a set of abnormal data, each of which may occur rather frequently during their daily work but which in just the combination in question may have been caused by a serious or dangerous failure that would have been considered very improbable in advance. This means that the operator is not allowed to trust his daily operating experience but has to base his decision on detailed knowledge of the functioning of the plant and its response to different types of failures."[19] Under new conditions heuristic skill can limit responsiveness.

Much the same problem appears in the maintenance area. The television repairers of twenty years ago (before solid-state electronics) forgot all their electrical circuit theory shortly after completing formal training. Instead, they catalogued patterns of failure according to a set of implicit heuristics. As products of particular repair experiences, the patterns reflected television failures only in a specific locale, and the heuristics, once entrenched as idiosyncratic mental habits, were very hard to modify. Whenever different viewing habits produced unexpected patterns of failure, repairers might be unable to fix the television set. Because their rules of thumb could not be generalized, they could not accommodate new conditions.

Today's control maintenance technicians in an industrial plant continue to depend on heuristics and simple diagnostics, but limits to this procedure are becoming apparent. When maintenance workers must fix digital circuitry for logic operations, they cannot simply trace the flow of an electrical signal down a main path. "Such circuitry has several information routes with common crossroads, and very often the signals look normal when judged individually but are present in faulty combinations or codings."[20] In this situation the search has to be "planned or deduced from an understanding of the internal functioning of the system and by employing a mental model of the specific system anatomy and functioning."[21] Since the importance of digital circuitry will increase with the spread of microprocessors, the maintenance task should be correspondingly more difficult.

Heuristics, as a set of mental habits specific to a particular system, may simplify synthetic reasoning under normal operating conditions. Such rules can, however, inhibit diagnosis in unexpected situations. Thus the skilled supervisor of a particular plant may prove unable to deal with emergencies. Research findings suggest that without theory operators will learn to gather just the amount of information they need to discriminate among the types of failure they have already experienced.[22]

Learning to Learn

Each time operators diagnose a novel situation, they become learners, reconstructing and reconfiguring their knowledge. To develop skill at learning, they must understand in what ways they become aware. Through the progressive integration of breadth, density, and depth, operators must learn to learn.[23] Although we know little about this process, available evidence suggests that three design elements are critical in shaping a learning environment.

Integrating Work and Planning

We often think of learning as separate from acting and performing. But this attitude can make training unproductive. Operators, technicians, and maintenance workers often forget basic theory because they have no need to remember it. When their desire for competence pushes them to try to understand some aspect of the process, they will apply their knowledge to real-life situations. There is some evidence that operators understand production processes more deeply when their planning horizon is extended—when they grasp broader and broader cycles of the production process. There is the cycle of immediate production—the flow of materials from input to final product; there is the cycle of changeover, in which new (but expected) inputs are introduced; and there is the cycle of product change, as new and unexpected products or processes are introduced. As workers develop a more inclusive view, they become increasingly aware of differences in the operational features of the production process. In assimilating this new information, workers not only learn more about the system as a whole; but by reevaluating their old concepts in light of the new ones—by relearning—they also become aware of the learning process itself.

Two pieces of evidence link the planning horizon to learning. One study suggests that chemical plant operators have a conservative control style: they respond to variations in the production process without considering the effects of their behavior on plant profitability. Their planning horizon does not link longer cycles of purchase and sale to short cycles of feedback-based control. Consequently, their behavior is fragmented into small, cautious, sequential steps.[24] In contrast, in a chemical plant that was designed to maximize learning, a computer provided economic data relating operators' control decisions to profit-and-loss consequences. With the instruments producing not only continuous technical data but continuous economic data as well, operators knew more about the production process as a whole.[25] Other design features can similarly expand workers' planning horizons. When the effects of a control decision are not immediately apparent, a separate control panel that simulates the altered process can help workers gauge the economic and technical consequences of their decision, making any necessary corrections before the real results are in.[26]

Sensory Data
Symbolic or mediated environments, which may engage only one or two of their five senses, can limit workers' reactivity. The capacity to orchestrate the senses into a heightened awareness of patterns of events is a great strength of the human operator as against the computer. For operators to develop good diagnostic skills in symbolically mediated environments, compensatory technical innovations should return "feeling" to the operators' experience. The technology for doing so is already in place. Some writers suggest that a worker using appropriate sensing technologies might connect her own senses directly to the plant so as to become a cybernetic extension.[27] Her body would shake with plant vibrations reduced electronically to a human scale, and she would feel warmer or cooler as the factory temperature changed. Pressure and sounds could be similarly transmitted. In a sort of "sound and light show" the worker would periodically contemplate the plant by integrating all her senses into the observation process. Through this total involvement she would learn about plant dynamics in a way that simple model development or visual observation could

never reproduce. Self-awareness would be crucial in distinguishing her own internal body cues from the messages of the plant.

The Allocation of Function

Most important, the allocation of responsibility between the controls or computer and the operator must be dynamic, based on the operator's learning needs as well as the performance requirements of the system.[28] In the literature on job design, analysts typically divide functions between the computer and the operator on the basis of their respective strengths: the computer performs the routine tasks, the operator the novel ones. But this leads to a cul-de-sac. The operators are so underutilized that they lose touch with the production process. They cannot be bored most of the time and still highly responsive at any given moment. The allocation of functions must be conceived with learning as well as performance in mind. As far as the needs of the team permit, individual operators should be able to choose when and how to engage in broader planning or become involved in the details of plant dynamics. To take full advantage of this flexible allocation system, operators must think of themselves as learners.

In the industrial conception the machine takes over human skills, rendering the operator superfluous. It is assumed that all new knowledge is introduced from outside the process. As we have seen, however, outside knowledge is increasingly insufficient and incomplete. With dynamic allocation the deskilling process is reversed. Machines extend workers' skill rather than replace it. As knowledge is incorporated into machines, workers can reinvolve themselves at a wider and more comprehensive level of production. Through a developmental process, machines and workers together increase the store of practical and theoretical knowledge.

Work and Consciousness

As we move from preindustrial to postindustrial conditions, the relationship between work and consciousness is dramatically transformed. The skilled craftsman of preindustrial society is absorbed in his tool: his sense of self is bound up with the tool's shape, power, and limitations. Habits result from the constraints imposed by the tool and from the problem-solving behavior of the craftsman. This gives the craft a very integrated and smooth quality but at the same time limits the craftsman to a particular tool or set of tools. Skill enhances capacity within the specialized context of a particular tool but inhibits overall flexibility.

In industrial society that paradigmatic semiskilled worker, the assembly-line operator, is chained to his tool, but his mind is free to wander. He retains the freedom to daydream, to explore his fantasies, wishes, and fears. His private sense of self, alienated from work, is self without substance.

The control-room operator of a postindustrial society brings his own awareness to consciousness by orchestrating fringe and selective attention into a unified process of pattern-seeking behavior. He retains a substantive self-concept as an agent in work, but only in developmental situations: that is, only when he controls the transitions between physical realities. In this process he experiences himself as one who adapts. Thus the coming to consciousness, as a fundamental element of postindustrial work, has two stages: the worker becomes more aware of his work environment, but he also begins to reflect self-consciously on his own actions and becomes aware of how he learns and develops. He learns who he is when he changes.

Developmental Work: The General Case

Expanding the Concept of Failure

In examining the case of the nuclear reactor as a model of the post-industrial integration of labor and technology, we have seen a new concept of work emerge. When second-order failures disrupt the error regulation of social and technical systems, prior experience may actually impede worker responsiveness. Moreover, in these genuinely novel situations the control systems themselves may fail. Stuck valves, flawed meters, and erroneous set points all become part of the accident. In these cases the workers must control the controls.

Is the nuclear reactor a good paradigm for this mode of work? Surely the consequences of failure are more far-reaching in a reactor than in other industrial plants. Might not the development in the latter of flexible workers and work teams be much less relevant?

There is little doubt that the nuclear plant and the chemical refinery are becoming the paradigmatic settings of postindustrial manufacturing. Industry depends increasingly on continuous-process technology and remote-control operation and supervision. To be sure, there are persistent differences among industries. Chemical operations and power production are inherently invisible processes, while steel rolling and machining are not. The worker, distanced from the former, must get information through mediating instruments; in the latter, the worker may still see, smell, hear, and feel the production process.

Yet even here there are changes. In integrated steel rolling, as we have seen, the operator sits in a pulpit and surveys the production process, controlling it with levers instead of through intimate contact. Similarly, machining may be technologically transformed from a mechanical to a chemical process if laser beams are used to evaporate metal to produce the required shape. Under these circumstances the worker will no doubt be removed from the machine to a control room. As one author notes, "There is a large element of centralized remote control in many modern processes."[1]

We must both refine and expand our definition of failure. Breakdowns will persist throughout the industry—boilers will burst, turbines and pipes will crack, electrical contacts will break. These failures can impose great pressures on work groups. But second-order failure often takes a more subtle form. Although local feedback devices may control many parts of a production process, they are still difficult to integrate into an overall framework of control through a master computer. Models of complex production processes must accommodate rapid changes in heat, chemical composition, speed, and materials.

I pointed out in chapter 6 that engineers in the cement industry still cannot successfully model the operation of the kiln. In the production of pulp for paper, operators must integrate the separate controls. In a pulp factory recently the subject of research, workers supervised the controls because different kinds of fibers (spruce, various hardwoods, and fir) had to pass through the same conversion equipment.[2] The operating rules of thumb had to accommodate varying inputs. Moreover, when the equipment was modified to cope with these variations, predesignated processes and production standards, based on blueprints of the machinery, became obsolete. A complex, informal system of work knowledge evolved in which the operators standing watch over the controls had to respond to changes in the conditions of production—that is, in the nature and composition of the raw materials. The operators were engaging in second-order work, the management of novelty, even as the machinery of production became more automated and the process became more continuous. Similarly, in a wax production plant the raw material used differed from the feed stock for which the machinery had been designed. Production suffered severely until management solicited worker observations, skills, and learning.[3] Thus changes at the boundaries of the production system, rather than changes within it, mobilized worker talent through problem-solving activity.

When workers participate in "quality circles" for problem solving they often demonstrate the value of shop-floor knowledge and expertise in improving production methods. In one factory disposal of the sludge in the tank of a grinder took an hour each day, until a worker discovered that when the tank was not replenished with water after a day of work the treatment went faster. Subsequent statistical studies by workers revealed that sludge time could be most reduced if the tank was filled to three-fifths of its prior level.

All of this suggests that continuous processes must be periodically adapted to changing conditions in the input (or input specifications). The sheer complexity of the mechanical-electrical processes and the continual modification of the technical equipment places developmental responsibilities on workers. We do not have to posit a series of extreme breakdowns or accidents to forecast the development of second-order work at the center of worker responsibility.

Failure emerges as a relative category. Short of breakdown, a production process "fails" when it does not meet output specifications; the degree of failure depends upon the quality of output required. In a paper mill variations in the growing and storing conditions of the timber used for pulp, if not compensated for in the conversion process, might reduce paper quality. The reduced quality would amount to failure, however, only if the output specifications were so tight that the paper produced was unacceptable. This mode of failure is clearly more complex than a simple breakdown or accident. It depends on the demands of users and the rate at which their needs change.

If we refine our concept of failure and breakdown so as to make failure a function of user needs and specifications, we may ask: Does the quality of output increasingly influence marketing and the profitability of enterprise? If so, this would support our contention that developmental work is becoming increasingly important to workers on the factory floor.

Finally, developmental work can become the model of labor if the variety of outputs increases. I have argued that with changes in raw materials workers must themselves modify the control systems so that the technical equipment can continue to produce acceptable products. Output variety involves similar issues. Output variety was formerly managed by simply adding new machinery or building an entirely new plant. Today's machinery is more flexible. Engineers can modify existing equipment by changing the controls, but this requires prior knowledge of the interactions within the control systems and between the controls and the raw materials. Thus product variety would again require developmental work as workers modified the equipment, the controls, and the protocols for integrating the controls. Is there evidence, or at least a theoretical rationale, for believing that the variety of input, the variety of output, and expectations as to quality are all increasing? If so, are managers, unions, and engineers encouraging workers to make developmental decisions?

The Olivetti Company
In an excellent case study Federico Buterra examines the relationship between changes in the Olivetti product mix from 1961 to 1971 and

the design of jobs and work groups in Olivetti factories.[4] Over this ten-year period "sales of office equipment like typewriters, . . . calculators, and office furniture dropped from 68.1 percent to 49 percent, while sales of more sophisticated products like accounting, data processing, and numerically controlled machines increased from 22.7 percent to 38.8 percent."[5] This shift reflected a change in the character, demands, and needs of Olivetti's customers. "The market for microcomputers, data processing systems, numerically controlled machines, and so on is more specialized, more differentiated in terms of performance requirements, more demanding in terms of technical quality, and above all subject to stronger competition. To be successful in these markets, the company must not only offer competitive prices and a very high quality; chiefly, it must also supply a wide range of products and constantly improve them to meet customer requirements."[6]

Buterra examines the ways in which management engineers and union officials cooperatively reorganized work in response to the pressure of this new market structure. First, the jobs of machinists who produced mechanical parts for calculators were enlarged, so that a worker planned and executed the production of a specific mechanical part and then inspected its quality. Formerly, in traditional machine-shop fashion, different workers handled setup, execution, and inspection—an effective design only when production specifications were loose. As quality standards tightened and the rate of production increased, more parts were rejected. Job designers appropriately reasoned that the divisions of machinists by grade and function prevented individual workers from accurately observing or correcting the sources of error. The jobs were consequently enlarged, not just to give the worker a "better" position, though this was an issue that interested both the unions and the designers. Engineers wished to design a machine interface through which workers could control unpredictable errors in execution and setup as well as uncontrollable variations in the quality of raw materials. As a result fewer defective parts were produced (the reduction ranged from one-sixth to one-third in the various sections of the workroom), supervisory labor cost less, and worker absenteeism fell.

A second change in job design concerned the assembly of accounting machines. Prior to the shifts in Olivetti's product market structure, accounting machines consisting of both mechanical and electronic parts were assembled in the traditional fashion. Workers sitting at stations along a manually paced roller belt fit and adjusted parts together according to clear geometrical relationships. Every five minutes or less, completed subpieces were pushed along the belt to the next stations. Although productive and efficient, this arrangement inhibited Olivetti's responsiveness to unpredictable market demands.

The [accounting] machine was a technically innovative product, very expensive, very sophisticated, intended for customers used to writing their own specifications for the product rather than buying what the manufacturer offered. Competition was severe. Obviously, market forecasts had been made, but they seemed less reliable than usual. Therefore production had to be flexible, that is, capable of making more or less than planned. The company could no longer do as it had with its mass-produced highly mechanical machines: expect the sales organization to sell more than the market demanded or back-order when demand was high.[7]

Olivetti developed a different assembly design to cope with quality-conscious users in a highly differentiated and changing market. Designers broke up assembly lines into cells or short lines, each of which produced one complete subassembly (such as the paper advance unit); each worker completed between one-fourth and one-sixth of the subassembly. Average cycle time, or time required to complete an assembly task, rose from five minutes to as much as fifty minutes. Individual workers inspected their work and supplied their own materials. Finally, at several assembly cells designed for training and experimental purposes, workers assembled the entire product in about two hours.

The conversion from a standard assembly line to a cellular design does not suddenly free workers from their drudgery or enable them to determine the conditions of their lives. On the other hand, workers appear to prefer these new arrangements. With a wider view of their activity workers can control the pace, intensity, and quality of their work.

The designers were more concerned with work-group and organizational flexibility than with worker satisfaction. Traditional assembly lines, although efficient, impose hidden indirect costs on the organization. Each time the product mix changes, engineers must redesign and recalculate work methods, standard pay rates, and new algorithms for "balancing the line." No worker, tool, or material may be kept idle for too long, despite the different cycle times of different subassembly tasks.[8] If engineers wish to experiment with methods for assembling a new product, they must improvise to find the space, workers, and equipment without interfering with the actual production process. In contrast, the modular or cellular process at Olivetti did not require extensive line-balancing calculations for the assembly of each product, since the interdependence among workers fell as cycle times rose. Experimental products could be assembled simply by adding another module or cell. To be sure, the decline in indirect and supervisory costs was partly balanced by new costs. Tools stayed idle longer, more workers had to be hired to increase output, and space had to be kept available for the addition of new modules. But these extra costs could be borne if the entire operation could respond to the changing demands of customers and the strengths of competitors.

These changes in the machine shop and the assembly room highlight the critical relationship between output variety and quality and work-group organization. Quality demands led to enlarged jobs for machinists, not "for their own good" but because only in this way could unpredictable errors and variations be observed and controlled. In the assembly room the demands for product variety, quality, and novelty led to enlarged assembly jobs because only in this way could errors be detected and workers prepared for new assembly processes.

The enlarged job gave assembly workers a wider view; more than that, by enlarging their skill and their repertoire of actions, the new job increased their capacity to adapt quickly to new assembly tasks. No product assembly system ever poses entirely new conditions; rather, workers face a new sequencing and integration of a set of modular tasks. A worker more familiar with these natural modules (the fitting together of entire components) is better able to link them together in new ways. Learning takes place as the worker decomposes tasks previously mastered into their natural subunits and then reorganizes these subunits around a new integrating process or task. In contrast, a worker accustomed to performing only a narrow mechanical task is not familiar with these modules and must be trained for a new subtask by the designer and the engineer—a process that costs the company considerable time and preparation. Thus the newly enlarged assembly job subtly redefines the worker: he is someone who not only contributes muscle power or fine motor skills to a task, but also contributes learning potential to a process of adaptation and change. In this sense the workers at Olivetti perform developmental work.

Fiat

Reports from other companies document similar processes over the last ten years. At Fiat, beginning in 1973, the market demand became uncertain, with sudden changes in quantity and quality. Managers decided to design "production facilities with the highest of flexibility. The objective is to be able to produce in the same shop and at the same time different car models, changing very quickly the mix of production. In addition the new production facilities must be capable of accommodating a new model with a minimum of time and cost required to perform the necessary adaptation."[9]

This commitment to flexibility led to several technical developments. To move parts between presses, engineers designed a transfer machine that could handle different parts of the same car or different car models through the resetting of specific parameters on the control system. The machine reduced the amount of heavy labor required, so that work became "increasingly concerned with controlling the production process."[10] Similarly, a "robogate" welding system was introduced through a series of cumulative technical improvements. Twenty-five robots, controlled by a central computer, spot-welded the car body as it was carried between work stations. The system could simultaneously weld two different models, thus "providing the plant with an unheard-of level of production flexibility and a consequent capability of meeting changing market requirements."[11] Finally, assembly was organized along lines similar to the Olivetti plan. Groups of workers were organized around subassembly tasks. A process computer directed the movement of a pallet between the work stations as the car was assembled. Subassembly was no longer tied to an unchanging line rate, and work teams could arrange their tasks to suit their own group needs for variety, pacing, and training. A computer mediated the exchange of information between terminals at the work stations. Such an arrangement expanded the skills of individual workers and the flexibility of the teams.

Citibank
Citibank also faced the problem of flexibility versus mass production.[12] The growth in the volume of business increased costs by 15 percent per year during the sixties, with more data-processing errors and larger backlogs of customer inquiry. Citibank reorganized its entire operation to increase the ability of its staff to respond to the specific needs of particular customers. A functional organization design, based on specialization and a large staff-to-line ratio, was replaced by a product-based design, and each clerk was made responsible for serving a particular set of customers. Data processing was decentralized, the clerks using minicomputers to obtain direct access to their customers' accounts. With the elimination of the traditional distinction between the front office and the back office, clerks could respond flexibly to the variety of specific customer needs.

The New Market Environment

How representative are the conditions that led Olivetti, Fiat, and Citibank to reshape their production processes? Equally important, how representative are the firms' responses? I believe that the shift to markets of greater variety and higher standards of quality affects all products and services. The available evidence supports this claim.

A recent analysis of the West German economy suggests that the number of new types of products has been rising exponentially since the mid-seventies, while the average demand per type of product has been falling.[13] A study of future market trends in the United States suggests that the mass markets of the industrial period are breaking up. In bread production, for instance, white bread output dropped 15 percent between 1972 and 1977, while production of specialty wheat varieties increased 62 percent.[14] Similarly, the fashions of the sixties reorganized the men's shirt industry so that the large shirtmaking companies had to invest in flexible automation technology to keep their costs down.[15] Finally, firms in advanced countries have shifted to a specialty products strategy in order to compete with firms using cheaper labor. Steelmakers, for example, are investing in mini-mills in which electric-arc furnaces and continuous-casting machines produce small runs of specialty steel.[16] Chemical companies have followed suit. Specialty products account for one-fifth of Du Pont's output, and management draws increasingly on the profits from mass-produced chemicals to upgrade the production of specialty chemicals.[17] In general, specialty grades in metal have increased dramatically in the last two decades, while specialty building materials, tools, medical products, and office equipment have so proliferated that a manufacturer can no longer rely on a single, slowly changing product with guaranteed name recognition and market.

This product variety is in turn reinforced by the breakup of clearly differentiated product markets. As the information capital content, or microcircuitry, of many familiar products—such as telephones, automobile engine parts, thermostats, watches, clocks, televisions, and radios—increases, the designs and uses of these products begin to overlap. Just as there is a growing divergence of products within any particular market, so there is an increasing convergence among product markets. Companies can no longer clearly distinguish their competitors on the basis of previous participation in a given market. The capture of the watch market by digital watch companies is representative. The information capital content of digital watches did not develop from watch-production technology per se, and thus traditional watchmakers could not anticipate the scope and strength of their unexpected competitors. The communication-information industries face a similar market disorganization. Until about 1970 analysts could clearly distinguish between information-processing companies on the one side and communication companies on the other. But as new products and services have integrated the two technologies, they have merged into one communication-information industry.

This diffusion of product markets and the creation of much more fuzzily defined "product systems," based on place of use (household, factory), are reinforced by the increasing flexibility of the capital stock itself. In the past a company might respond to shifts in material supplies or market demands by adjusting the size of its staff or its facilities. With a fixed stock of worker skills and capital the firm could respond only by changing its human and capital assets. The discontinuous nature of this response imposed serious costs that limited the company's ability to enter new markets or compete with new processes. Today, however, as the capital stock itself becomes more easily modifiable and as workers enlarge their skills and learning potential, the cost of such adjustments is falling. We have seen that such industries as machine assembly, chemical processing, and plastic forming increasingly display this generic flexibility. The machinery no longer has a fixed cycle of action: workers and designers can modify the controls so that a machine can perform new functions. Consequently, the barriers to markets are falling, and the diffusion of processes is increasing. Assets, skills, information, protocols, and procedures can be transformed rather than scrapped, because they already display generic flexibility. This change is authentically developmental.

The same forces that are increasing product variety are also raising quality standards. Always defined with respect to use, standards become more exacting as users' demands become more specific, various, and stringent. Thus product quality, like variety, increasingly determines profitability.

This trend is most clear in computer sales. Manufacturers are discovering that theirs is in fact a service industry. They cannot sell computers unless users can clearly identify the computer functions that meet their information needs. To be sure, standard functions, such as payroll and billing, can be automated and sold on a mass basis, but with the declining cost of information processing, customers are buying hardware to develop more complex systems that closely match particular management and production requirements. Thus the computer companies must increasingly consult with potential users, map their management and decision-making structures, and then arrange for each user an open-ended hardware-software package that best matches its current and projected needs. Thorough consultation is essential, not only because poor information systems can seriously undermine management, but also because call-backs and retraining of users are very expensive.

The increase in the information capital content of many intermediate and final products also leads to more exacting user needs. Such goods as electronic thermostats, alarm systems, and the fuel injection systems of automobiles function as control elements in the larger systems of households and cars. Small errors in a control system can have large impacts on the system in which it is embedded. Small deviations in the circuitry that controls fuel injection can have a sizable effect on the rate of gasoline consumption. Control system errors have this multiplicative impact because they set the reference points (temperature, pressure, viscosity) for other components of the system. Biases in the reference points will lead to systematic overuse or underuse of certain resources, while fluctuations in the reference point will lead to poor and unpredictable performance. A control system error is like a brain dysfunction. Small errors in the brain will lead to large errors in body functioning.

The rise in user standards demonstrates that the cost and availability of products are less important than their quality and relevance. This is evident in the service sector, where hospitals often have too many beds, and schools, though they command large human and physical resources, cannot teach children. Similarly, the automobile companies are not constrained by the number of automobiles they can produce or the productivity of automobile workers, but rather by the lack of fit between the car and the general ecology of transportation, that is, the transportation system in its widest features.

The new emphasis on quality as a critical determinant of market demand signals a shift to a service economy. A consumer is satisfied with a service when care is shown in its delivery—that is, when the service has been made to fit his particular needs. Thus as quality standards for goods as well as services rise, the manufacturing sector must pay increasing attention to the multidimensional ecology of use that gives economic value to its products. In the past a company may have successfully carved out its own market on the basis of effective "impression management." But as product boundaries become diffuse and product systems emerge that bring separate industries under the same ecology of use, these strategies prove ineffective. Thus the company is progressively exposed to a more varied, more exacting market.

Within this framework it must develop a generic flexibility to respond to unpredictable variations and changes in setting.

Postindustrial development is thus self-perpetuating. Cybernetic technology increases the generic flexibility of the capital stock; the information capital content of goods creates market uncertainty; and more exacting standards, combined with the wealth-creating potential of the new technologies, create market structures in which quality demands and standards are the most important determinant of profitability. These trends all reinforce one another. The flexibility of the capital stock (and of the labor pool attached to it) supports, and is supported by, the greater variety of market conditions.

In Sum

The cases described above outline the conditions that can lead managers to redesign jobs in ways that increase the learning potential and error-detecting capabilities of workers. Are Olivetti's, Fiat's, and Citibank's responses to these conditions representative? In the short run, companies will find many ways to cope. Those that can still protect their markets through commercial or political strategies may do so. Large corporations may buy up companies in other markets to accommodate shifts in product demand, or they may protect profits by building in flexibility not only at the most basic level—that of the machinery and the workers—but also at the upper levels, by integrating many companies through a unified financial system and a single balance sheet. In this way the misfortunes of one division are likely to be offset by the success of another. Many American companies are clearly pursuing just such a strategy, though much evidence suggests a resulting decline in the flexibility and adaptability of our entire economy. Nonetheless, there are indications that the Olivetti strategy is finding its adherents as well. By examining instances of the latter we can see how the new technologies and concepts of work based on learning are jointly creating new work settings.

III

People at Work

11

Integrating Work and Learning

As factories invest in and develop the new technologies, managers must develop job and organizational designs that facilitate learning. The nuclear power industry has been backward here. A worker who will be supervising a cybernetic machine should be trained on a simulator in which a computer model mimics the operation of the machine. Interacting with the simulator, the worker will learn how one decision can set off a train of events. Yet the managers of General Public Utilities, which owns the Three Mile Island plant, chose not to spend the five million dollars needed to design and develop such an interactive simulator. Rather, they instructed workers on "single-fault" failures—how to respond to an open valve or a rise in temperature—with training materials and devices that were not interactive.[1] Real accidents, however, often proceed through a chain of events—a set of interdependent failures, with one failure increasing the probability that another will occur—and in interaction with workers.

At present there are only about twenty simulators around the country. Workers are reluctant to spend time at simulator centers, away from their families, and managers are equally reluctant to lose their workers for extended periods. Most training, therefore, is conducted at the plant, either by utility officials or by vendors who sell training packages to the companies. The utility trainers are usually excellent senior operators but not necessarily good teachers. Their courses are most often geared to the qualification examinations devised by the Nuclear Regulatory Commission (NRC). The aim of these tests is not to deepen workers' understanding of the physical, chemical, and systems aspects of the reactor process, but to test their competence at performing rote tasks and their familiarity with emergency procedures.[2]

The licensing examiners are often consultants assigned by the NRC to the utility. Although some experts have criticized the examinations as too easy, both the unions and management are reluctant to see workers disqualified. Examiners, plant officials, and union officials may unconsciously collude to help less than competent workers pass the tests. One veteran trainer estimates that "one out of twenty operators he meets scares him," and that the "required performance level for nuclear power operators is less stringent than the requirements for a commercial pilot's license."[3]

Technical complexity and safety requirements have compelled the NRC and the utilities to develop a system in which worker training is a central element in managing production. In this the nuclear power industry contrasts with traditional industry, in which semiskilled workers learn informally, if at all, and skilled workers learn mostly in trade schools or through apprenticeships. Yet the utilities have designed training systems that all too often confine the learning process to narrow institutional needs and rhythms. This training cannot meet the deeper requirements for expanded learning or equip workers to cope with the unexpected. It is likely that this discrepancy between training systems and learning needs will grow over the next decade. Until recently the utilities have drawn heavily on the intensive training provided by the nuclear navy, where the student-to-teacher ratio approaches one-to-one. In 1970, forty-five percent of all licensed operators had had such naval experience.[4] But as the number of plants increased, there were fewer nuclear navy people to staff them. In plants that went into operation between 1977 and 1980 only 11 percent of the workers came from the navy, while in plants put into operation ten years earlier 37 percent did.[5] Thus in the future the utilities will not be able to build upon the education already provided by the navy. Instead they must revise and expand the ways in which they train their workers.

Yet the utilities have been highly reluctant to encourage a problem-solving mentality in their workforce. Their tradition of management and control makes it difficult for them to consider new uses for workers' skills and intelligence. They wish to hire and train workers who will solve problems but not challenge management. As one training coordinator put it, "We don't want an engineer who would be too inquisitive, who would go around analyzing designs instead of operating the system."[6]

The New Setting

Designs for training in nuclear power plants, as in traditional settings, separate learning from work. Yet over the past decade companies in

North America have designed about five hundred factories, often referred to as "sociotechnical" factories, that integrate work and learning. While they vary in design and structure, they share a number of features, summarized here.

1. Workers are paid salaries, not wages. A salary is determined not by the worker's performance but by how much he or she has learned. The pay-for-learning system is organized on the basis of skill clusters. At a food factory that I will call "Family Foods," the base salary in 1979 was $850 per month for workers who had demonstrated competence in one operating cluster, such as roasting. They could rise to a salary of $2000 per month by demonstrating competence in all the operations of the plant, including quality control and maintenance.[7] Similarly, a chemical plant instituted a regular progression requiring "the acquisition of skill in six process areas and two other skill areas to reach the top and a laboratory progression requiring demonstrable skill "in three process areas and thirty-six lab modules to reach the top rate."[8]

Most plants organize pay clusters on the basis of the technical integrity of particular kinds of work. For example, there will be separate clusters for line, maintenance, warehouse, and laboratory work. While workers are encouraged to develop competence in all these clusters, most will "major" in a particular area and "minor" in others.

2. The workers are organized into teams based on natural segments of the production process, such as packaging, processing, and laboratory work. Such teams reflect the technical interdependencies among tasks in a particular part of the plant. Teams in most factories provide a range of roles through which members can develop new technical and social skills. A worker may be a materials scheduler, a work assigner, a trainer, a financial coordinator managing the team's budget, a health and safety coordinator, a recorder, or the team's representative on a committee studying social-system issues throughout the plant.

Team members occupy these roles while simultaneously managing production and maintaining machines. Thus, unlike workers in traditional factory settings, they both regulate the social system of intrateam and interteam relationships and sustain the factory's technical operations. Indeed, some workers may spend the bulk of their time in management or administrative roles. At a compressor plant one materials scheduler on a process team spent 80 percent of her time controlling inventory and only 20 percent on the line.[9] Workers spend time at team meetings, too: at a chemical plant teams took off one paid day every nine weeks to review team problems.[10]

3. Workers train one another and rotate through the teams to learn the full complement of skills. Since particular teams are limited to particular parts of the plant, a team member must temporarily transfer to other teams to develop new skills. This means that workers and their teams negotiate temporary transfer arrangements with others. If, for example, a worker wishes to transfer temporarily from the line to the warehouse, but the warehouse team is not accepting transfers in the coming month (perhaps it is experiencing interpersonal conflicts that would only be exacerbated by the presence of transfers), the line team may negotiate with the warehouse team for a guaranteed right to transfer in a month's time. At Family Food the arrangement was that two workers could sign a contract specifying when they would exchange places and when they would return to their home teams.[11]

Unlike workers in traditional factory settings—where skills and privileges are intimately connected—members of a line team, for example, want to train a transfer from the warehouse, because they in turn expect to be trained when they transfer to the warehouse. Workers may be trained on the job by any others who know the work, but most often there is one worker on the team who orients new members and sits on a general committee supervised by the training coordinator for the whole plant.

4. Workers evaluate one another for pay raises. At one paper plant the person was interviewed by the training coordinator, who then administered two tests. The first was a written test concerning plant and equipment operation, the second an on-line demonstration of skills and understanding. The company supplied substantial written material to aid in the preparation. If the results were satisfactory, the training coordinator would meet with the other members of the unit to discuss a predetermined set of factors (e.g., attitude, safety, conscientiousness, housekeeping skills, etc.). This group would then decide on the worker's abilities and skills. If satisfactory, the employee would be assigned skill points, which were correlated with salary.[12]

At another paper plant the team members and the team facilitator together evaluated each worker. The facilitator's final decision was based on the team's recommendation. In some plants workers who are dissatisfied with their evaluations can appeal to a central review board. Thus the evaluation process is subject to checks and balances that promote fairness.

5. The role of the first-line supervisor is changing. In some plants the supervisor acts as a facilitator or coordinator assigned to a particular team, while in others he is responsible for a functional area, such as training or health and safety.

6. An elaborate system of committees and task forces manages the plant. Some of these focus on particular projects and problems, such as training and plant expansion, while others review the entire design. In one plant a group with the latter function is called the "social system committee"; in another, the "norm review board."

7. Few of these plants are unionized. Many were built from scratch in regions where unions are weak. Unions are often reluctant to participate in the design and operation of such plants. They fear that collective bargaining will be weakened—that workers will identify more with the teams, the plant social system, and management than with the union.

The relationships within and between teams produce an organizational structure that is far more complex than the simple hierarchy of the traditional factory. There are five components to such a design: the teams, the project and standing committees, the support units, the supervisors, and, overseeing the rest, the social system committee. The latter is particularly important since it enables workers and managers jointly to assess the functioning and design of the plant as a whole. Its absence at a General Foods dog food plant contributed to a climate of stagnation and stalemate after six years of development.[13] Another approach has been taken at a compressor plant, where workers and managers assess plant functioning on a more informal basis through fifteen small review boards.[14]

Finally, this design places members in "matrixed" or cross-functional roles. For example, the team coordinator for purchasing interacts with the purchasing supervisor, coordinators from other teams, his own team, and the purchasing department. He develops a complex understanding of plant operations.

The Roots of the Design

The proximate cause for the emergence of these advanced work settings is the confluence of cultural change and economic crisis that is reshaping industrial structure and industrial relations everywhere. Particularly over the last decade, a growing number of managers, union leaders, government officials, academics, and consultants have argued that jobs must be designed to enhance the "quality of work life"—to meet emerging cultural aspirations for satisfying work—as well as to increase productivity. But the roots of the advanced work designs may be found in certain theoretical and practical discoveries made at the end of World War II.

The influential sociotechnical conception of work design was developed at the Tavistock Institute in London. Researchers led by Eric Trist incorporated concepts from anthropology, psychoanalysis, and systems theory.[15] From anthropology these sociotechnical theorists drew methods for mapping out the network of relationships among the material and symbolic elements of a system in which social structure and technique were interrelated. Psychoanalytic insights offered clues for interpreting the interactions among members of small groups or teams. Systems theory indicated that a group could regulate itself without supervision by using feedback on its performance.[16] Finally, Kurt Lewin's field theory, a branch of systems theory, taught that people are motivated to complete tasks—are pulled on by the structure of a task to seek a sense of closure.[17] These three traditions converged to create the concept of the self-regulating work team, completing whole tasks in a work system shaped but not determined by technological parameters and in a social system shaped by psychosocial group processes.

Sociotechnical theorists stressed the importance of informal group processes. While Elton Mayo had acknowledged the significant role of the informal group in his Hawthorne studies, he had viewed it as a system of relationships ancillary to the work system. The sociotechnical understanding of the team turned this conception upside down. The informal group process was the vehicle for organizing and regulating the work. Taylorist job design masked the role of the informal group by impoverishing its processes and fragmenting jobs; the supervisor became necessary.

For example, assembly-line designers calculated flow rates and work methods on the basis of averages. They assumed that on the average,

different operators work at the same pace, [and] an individual operator works at the same pace throughout the shift; . . . on the average, learning on the job can be ignored; on the average, variations in parts and equipment can be ignored. . . . But one thing is pretty sure: at any time on a line it is most improbable that all aspects are operating at their average values. Typically, something is always nonaverage, wrong; and when one thing is wrong so are half a dozen other things."[18]

Since group processes are almost completely disrupted or blocked on assembly lines, the foreman was expected to control those deviations from the average that eluded the methods engineer. He moved constantly about, supervising the work, performing necessary tasks as lapses occurred in the work flow, and pushing workers to produce at a predetermined speed. Thus the foreman's function hid the potential role of the informal work group from view. The foreman's role was self-reinforcing: workers took their job descriptions as the maximal rather than the minimal specification of duties, forcing the foreman to take all the initiative when contingencies and errors arose in the flow of work. Under these conditions workers often withheld critical information from management.

Two Experiments

In the early postwar years, Eric Trist and his colleagues applied their theory in field settings. They demonstrated that coal mining work was most productively managed when workers were organized into self-regulating work teams rather than divided up in narrow jobs.[19] They showed that the percent of coal won from each daily cut went up, that the amount of ancillary work at the face (preparation, cleanup) went down, and that the need for reinforcement fell. All this happened as the number of specialized task groups decreased and job variation for individual workers rose.

These results were confirmed in another setting when a co-worker of Trist's, A. K. Rice, redesigned an Indian textile mill.[20] Although engineers had installed automatic looms and assigned work load on the basis of rigorous time and motion studies, efficiency was low and damage to the looms was high. Rice examined the production system and discovered that the jobs were very specialized and highly inter-dependent, creating a crisscross of traffic among looms and jobs that made it difficult for workers to coordinate their work effectively. Rice reorganized the work system so that a single team operated and main-tained a single bank of looms, and all teams reported to a single shift supervisor. Efficiency rose and loom damage fell. Again, a job design based on the principle of specialization had produced a counter-productive production system.

Do These Plants Work?

Sociotechnical factories can be evaluated as approaches to the problem of productivity, as attempts to improve the quality of work life, and as experiments in industrial democracy. From none of these viewpoints, however, can one clearly perceive the meaning of these settings. Such factories are not necessarily more productive than traditional ones; rather, they are more flexible and can adapt more quickly to changing market demands and changing technologies. These factories most often do not give workers a share in company profits, nor are they organized to increase the influence of worker representatives in the formulation of broad company policy: they cannot, then, be considered attempts to develop the practice of industrial democracy. Finally, if we see these experiments as part of the "human relations" tradition, as attempts to make workers feel better about their work and to motivate them, we neglect the particular appropriateness of the new work forms to a postindustrial technology. This last point is most important for the present analysis, which seeks to describe what happens in these factories when managers and workers try to integrate work and learning.

I know of no systematic study comparing the long-term performance of these plants with that of conventional ones. Case studies and my own interviews with managerial and supervisory staff suggest that these plants produce a higher-quality product than do conventional factories, while remaining profitable. Their rapid growth over the last decade, from one in 1970 to about five hundred today, suggests that companies find them profitable.

In the absence of longitudinal research based on rigorous experimental designs it is important to dig deeper into the history and functioning of these settings, to see how their designs are implemented. How do workers and managers in such plants behave? People who work in sociotechnical settings face new dilemmas—the following, for example:

1. What happens if team members promote a worker who is not competent? They may do this if they are afraid of the conflicts that may ensue if they don't promote their teammate.

2. What happens when a worker reaches the top rate of the plant? How does he advance? How can he earn more money? Will he stop teaching others if he is no longer learning himself?

3. If a machine breaks down, is the team's supervisor responsible? Should he have stepped in and prevented the failure? Or should he let team members make their own mistakes so that they can learn?

To understand these issues, it is helpful to study the ways in which such sociotechnical settings fail to integrate work and learning. What typical strains develop that lead managers to limit worker learning and team autonomy? What conflicts arise among team members that limit the team's ability to control the production process? Just as we study the failure modes of technical systems in order to design better ones, so it is useful to study the failure modes of these factories. In so doing, we can understand the historically and culturally unique dilemmas and pressures that workers and managers in these plants face.

The Sample

In order to clarify the design and functioning of sociotechnical settings, I interviewed personnel from twelve plants, a warehouse, and an office, and I spoke with consultants who worked with managers and workers in eight other settings. I also consulted published and unpublished documents. Table 1 summarizes the outputs and technologies of the different settings. Where a factory has been described under its real name in published work, I have used the real name here; most have fictional names.

Twelve of the settings, all factories, are full sociotechnical designs; that is, workers are organized into teams, are paid a salary on the basis of how much they learn, and are not directly supervised by foremen. All of the full design settings but one were established in or after 1975. The General Foods plant was established in 1970.

A total of twenty-four interviews were conducted, seventeen of them with personnel from the thirteen factories having full sociotechnical designs. Twenty-two of the twenty-four interviews were conducted with supervisory staff or consulting personnel, and only two with workers. This represents a potential bias in the results. Assessment of published and unpublished reports on these settings, however, suggests that supervisor reports correspond very closely with worker reports. Published and unpublished documents on experiences at Fall Mills and Sound, Inc., were particularly helpful in understanding worker responses in sociotechnical settings.

Plant Technology and History
No systematic study was made of the reasons for establishing these settings. Casual observation suggests that three factors jointly determine the likelihood that such designs will be established: senior managers who are highly motivated to set up innovative work systems, the building of a new plant, and an advanced continuous-process technology.

Technology does not determine whether such settings are established, but it does play a facilitating or enabling role. Of the twelve full designs, nine have continuous-process technologies and two have partly continuous. The two chemical plants, the paper plant, the paper products plant, and the GM battery plant were based on new processes. The experience at Fall Mills seems typical: "The speed with which these [gluing] operations are performed requires careful monitoring of finely tuned electrohydraulic machinery. Tolerance limits are so fine and

Table 1

Name	Plant	Technology
GM Battery F/NU	Batteries	Continuous process
Family Food F/NU	Food	Continuous process
Fall Mills F/NU	Paper products	Continuous process
Scott Compressor F/NU	Machining	Partly continuous
Fuel, Inc. P/U	Gas	Continuous process
Reading Meat F/U	Meat	Mechanical
Office, Inc. P/U	Data/Typing	Computer work stations
Smart Chem P/NU	Plastics	Continuous process
New Chem P/NU	Plastics	Continuous process
Big Chem F/U	Plastics	Continuous process
Little Chem F/NU	Plastics	Continuous process
Trucking, Inc. F/U	Warehousing	Manual, Assembly
Farm, Inc. P/NU	Farm products	Mechanical/partly continuous
South Mack F/NU	Ball bearings	Mechanical/partly continuous
City Bureau Springfield, Ohio P/U	Service	Craft
Household, Inc. P/U	Consumer products	Varied
Sound, Inc. P/NU	Electronic components	Assembly
Oil, Inc. P/U	Oil	Continuous process
General Foods F/NU	Pet Food	Continuous process
Paper, Inc. (1) P/NU	Paper	Continuous process
Paper, Inc. (2) F/U	Paper	Continuous process
Electronics, Inc. F/NU	Circuits	Continuous process

Code: F = Full sociotechnical design   P = Partial sociotechnical design
    U = Union plant   NU = Non-Union plant

interdependent that sources of machine stoppage and low-quality product often seem difficult to identify and impossible to eliminate entirely."[22] During Fall Mills' first year of operation technological change was so rapid that "even before the last five departments were fully operational, some of the first department's production lines were gutted and newer-model machines were installed."[23] On one typical shift workers made seventeen "remedial" decisions (small control decisions that did not entail major diagnoses of machine failure), resulting in a total downtime of sixty-nine minutes, or more than 12 percent of the total shift.[24]

The computer system at Big Chem was specifically designed to facilitate worker learning. It did not control all the feedback loops, because engineers could not fully represent the dynamics of the system in mathematical terms. In other plants based on similar technology, attempts to maximize computer control had led to significant under-production. The computer was therefore designed to facilitate learning. "The computer answers queries put to it by the operating personnel regarding the short-run effects of variables at various control levels, but decisions are made by the operators. Operating personnel are provided with technical calculations and economic data, conventionally only available to technical staff, that support learning and self-regulation. In this manner operator learning is enhanced."[25] At Smart Chem the process engineers did not automate the control system. Managers wanted the workers to learn the process dynamics by directly controlling the machinery.

Managers are likely to introduce advanced work designs in advanced technology settings for two reasons. First, it is often easier to develop and implement new job designs in new plants, and new plants are most often supplied with the newest machines. Second, supervision becomes less directive and more facilitative in such settings, even in the absence of deliberate sociotechnical job design. This happens because the foreman need not and cannot push operators to work harder. The pace of production is determined by the machine; workers must solve control problems.

The new technologies thus provide a more receptive framework than do traditional technologies for the development of sociotechnical designs. Moreover, such designs are reinforced by subsequent developments. The flexibility of the technology encourages managers to make changes when it seems profitable to do so, while the workers themselves, used to learning, more readily accept such changes. Thus advanced technology, changing technology, and a job design based on learning reinforce one another.

12

Developmental Tension

Failure Modes

Workers and managers at Big Chem faced a critical juncture in the first winter of the plant's operation. The plant was highly automated, the technology was new, and workers played a critical role in controlling the production process. Yet the workers could not master the machinery. Top management, afraid of year-end losses, installed temporary assistant coordinators to supervise the work teams. The workers were angry: they had hoped to develop a team system in which workers were truly self-managing. Most had assumed that the team coordinators appointed by management would play an increasingly subsidiary role. Yet management was now appointing assistant coordinators, creating a new level of supervision in the plant. Here are three assessments—a worker's, a manager's, and a consulting team's—of the situation at Big Chem.

This company has serious problems with the QWL [quality-of-work-life] concept. They grafted a radical organization onto an old one, and we ended up with panic. In the end top management, despite their support for QWL, had old ideas about supervision and how you manage a start-up.

   You have to realize one thing. This is a solid plastics technology, the first of its kind in North America. Sixty percent of the personnel did not know the equipment, and health and safety training was not geared to the specifics of the technology. When the feed backs up it backs up quickly. There is little time to search for problems. When the process is shut down, you can't take the plastic out with a machine. You have to take it out by hand, and it's messy junk. There we were in the middle of the winter, up to our knees in the junk because we didn't know how to manage the technology. Plant management panicked. They were watching the bottom line fall away from them, and they appointed temporary assistant coordinators to each of the autonomous teams. They covered their ass that way and got the head office off their backs. But it was a unilateral decision: they pulled us out of the plant one day and announced it. The decision should have been made collaboratively. We complained about the way they did it. Why were they being so paternalistic and authoritarian? I guess that we had a misconception: we thought we were on our way to autonomous work groups.

I'll admit that in the end more than half of the workers supported the new position of assistant coordinator. But partly the company co-opted the issue by setting up a study to review the assistant coordinator role. They called the role a temporary one and denied that they were responding to union pressure. We felt that with twelve people on the plastics side of the plant the role of assistant coordinator could have been divided up among the team members. We would have come up with the skills; we knew things had to be done. There was another option: we could have nominated a subcommittee of three or four people to fill the coordinator role.

Management filled the assistant coordinator positions. Thirty-one workers applied for the jobs. Five out of the six coordinators welcomed them; one quit, and all [the workers] felt blamed. We know what kind of people applied for the job: they were the types who want to be chairman of the board of directors. We tell them that it's okay, but their future lies out of the plant, not in it.

I agree that the teams stabilized when the assistant positions were filled. Before that, when an informal leader would emerge, he was cut off at the head. People got too comfortable with committees. The assistant coordinator had a unifying effect. You had two people now. It wasn't hard for the coordinator to be around and be available. Before that he was chasing his shadow. The teams settled down. (interview with a worker, 1982)

This plant is not moving toward a traditional supervision system. The assistant coordinators are helping hands. They were team members and were respected. This is a huge complex with a new technology, and it is difficult to manage. People didn't know the technology and were asking for help. The administrative workload for the coordinators was enormous. They were dealing with technical problems, communication problems, and with upper-management jitters. They felt isolated and needed support. Some didn't know how to let go: they had come from traditional settings. The teams needed more boundaries between themselves and the rest of the organization—two coordinators instead of one, more team structure, and more training in meeting skills. I've learned that you have to design autonomous work groups; they don't just happen. The team workers are concerned that they are not doing as good a job as they can. They worry about the fact that the skills are being watered down; they are afraid of being pressured from without; they worry about the system being removed. I know that there are disgruntled members. They would say that the coordinator does not involve them in decisions, [that] team members should not be allowed to discipline one another, and that the progression system isn't good—people are rising too slowly. But I have talked to all the workers. They all want authority. (interview with a manager, 1982)

The technology and technical system chosen called for continuous processing using various vessels, reactors, and remote control of chemical reactions that could be physically dangerous if not properly done. Equipment and instrumentation costs required a massive dose of capital investment, amounting to approximately two million dollars per employee. The characteristics of the technology, the size of the investment, and the small number of employees (150) led to the recognition that economic success would depend on the willingness and dedication of this small number of employees. The recognition that higher levels of technology, frequently accompanied by very large capital expenditures, increases the dependence of organizations on their workers, rather than the opposite as predicted by engineers, was crucial to the design process. This awareness, plus a prior history of examining and searching for work relationships that would reflect the high level of responsibilities placed on workers, led to dissatisfaction with the bureaucratic scientific-management structures prevalent throughout the industry. (report by consultants on the plant design process)[1]

The worker and the plant manager reveal some of the ambiguities and difficulties that the members of sociotechnical factories face. The worker is angry because the teams were not allowed to become autonomous, yet she realizes that the plant was dealing with a technical and financial crisis. Moreover, she admits that the teams stabilized after the assistant coordinators were installed. She points out that another solution was available—team members could have divided up the role of the assistant coordinator—yet notes later that "people got too comfortable with committees."

The manager knows that the workers are not irresponsible. They worry that they are not skilled enough and fear that they have not confronted they complexity of the tasks they face. They fear that the work-group system will be "removed," that coordinators do not support team development and will not involve them in decisions. Coordinators as well as workers need time and support in order to learn how to function in these settings. Coming from traditional settings, coordinators too often take control of the teams and do not allow team members to learn and develop. Yet the manager has learned that autonomous work groups don't "just happen"; they must be "designed." More structure and organization are needed: all the workers "want authority."

But the worker is suspicious. Even if the team system stabilized after the assistant coordinator was appointed, why had management been so "paternalistic and authoritarian"? Why did they make the decision unilaterally? The process violated the collaborative spirit of the plant. Had the decision been made collaboratively, workers might have accepted more readily the new layer of management. Workers as well as managers worried that the teams lacked technical competence. If the plant's regression to a more traditional management system was only temporary, then the increase in the level of organization and structure might prove beneficial. Workers would lose some autonomy now but be more capable of managing themselves in the future. If the regression was not temporary, however—if management was simply co-opting worker anger with a study—then the added layer of supervision was a sign that workers had lost their chance to become fully autonomous.

Such ambiguities and dilemmas affect many sociotechnical plants. For example, South Mack was located in a Sun Belt state where local workers lacked sophisticated machining skills. Yet the turning and grinding lines "used equipment that is recognized in the industry as very difficult to keep in tolerance at high speeds."[2] The workers had been trained to set up and operate basic machining equipment but were unable to diagnose problems and make adjustments when the machinery was producing at high volume. Fearing that the plant would prove unprofitable, management promoted several workers to the position of "technical assistants." Such assistants were paid top wages and were expected to troubleshoot throughout the plant. After the decision was announced, "all hell broke loose."[3] Workers were angry over management's violation of the pay-for-learning system. The assistants had not advanced up the ladder, and many workers felt that their competence did not warrant top-grade pay. Moreover, as at Big Chem, workers resented the fact that the decision had been made unilaterally. Management action violated the premise that "people would have the opportunity to influence those decisions that would affect them."[4] Ultimately, management appointed worker specialists to set up the grinding and turning machines, thus eliminating these tasks from the rotation system.[5]

These cases highlight some of the problems managers and workers face in balancing team and individual development with plant performance. The short-term exigencies of the technical system may limit the workers' long-term opportunities for learning. Management may undermine the pay-for-learning system, may impose new layers of authority on the team, or may reintroduce job specialization and fixed job categories.

Often, however, the technical system may not prove challenging enough. At the General Foods plant, for example, levels of commitment to the system began to erode after four years. The plant was operating smoothly and management had not introduced any new machinery or products. Walton argues that a "human resources surplus" grew up during this period.[6] Workers felt complacent. As a result cooperation between shifts, helping behavior within and between teams, and confidence in General Foods all fell. Workers were overly skilled for the technical challenges they faced.

This suggests that if sociotechnical factories are to function effectively, if they are to promote both individual learning and group performance, they must find some optimal level of what I call "developmental tension." I use the word "tension" to express the ideas of uncertainty and contingency. Because sociotechnical settings encourage member learning and development, the number and range of rules, procedures, and authority relationships governing behavior are smaller than in conventional settings. The resulting uncertainty can create tension and anxiety, but it is this very tension that provides the space and the sanction for creative and innovative behavior.

The level of developmental tension is determined by several inter-related factors. As the degree of novelty in factory life increases (management invests in new and more complex machinery, for instance), so does developmental tension. But as worker skill increases, these same events may prove insufficiently challenging. Yet if management takes greater control or introduces greater structure into the plant system (by imposing task specialization, for example), even very difficult tasks, such as introducing a new product to the line, may not challenge workers. This complex of relationships can be summarized in the following heuristic formula:

Developmental Tension = Novelty/(Skill × Structure)

Sociotechnical plants can fail to integrate performance and learning in three distinct ways. First, the level of novelty is too high relative to worker skill, and management introduces a new level of structure or organization that permanently reduces future opportunities for learning. Second, the level of novelty is too high and management does not introduce either structures or training programs to help workers get over the hump. Continuing failure demoralizes workers; and when management finally responds, it is too late. Workers lose faith in their own competence, and their commitment to the team and the learning concept is permanently reduced. Third, the level of novelty is too low; workers and management lose interest and fall into a procedural routine that reduces future opportunity for learning.

At Fall Mills the skill clusters were divided into process, warehousing, and maintenance work. Workers had trouble mastering the technology.

The automated machinery was complex and there was a high rate of technical change. "Even before the last five departments were fully operational, some of the first department's production lines were gutted and newer-model machines were installed."[7] Workers saw that the process operators were having a hard time managing the machinery, while warehouse work, which included the operation of a computer for inventory control, seemed easier.

Several workers who began their rotation on the warehouse team lacked confidence in their mechanical skill. They argued that the warehouse skill cluster should be treated as a separate career path, that they had "a right not to move further." Line workers and supervisors, who were dealing with complicated new machines, found it burdensome to train transfers from the warehouse. Management therefore agreed to split the two clusters so that workers from the warehouse would not rotate to the line.[8] The scope of potential worker learning was thus permanently reduced. To cope with excessive developmental tension, managers and workers permanently increased the level of structure or organization by creating specialized job tracks.

At South Mack managers consistently underestimated workers' need for both technical and social leadership. They failed to introduce procedures, skill training, and task specialization to help workers cope. Management discovered that workers in the Sun Belt lacked the machining skills found in a northeastern community. They took corrective steps, such as bringing in technical support staff, but by their own admission these were far too small. Walton argues that they became the victims of their own "wishful thinking," hoping that a self-managing system rooted in deep worker commitment to the plant would produce solutions to difficult technical problems in the absence of technical leadership.[9] When a new plant manager was finally brought in, he stabilized the plant by asserting such leadership. He established a separate quality-control unit, introduced a preventive-maintenance program, took setup work in grinding and turning out of the task rotation system, increased the ratio of supervisors to workers; "he made the plant spotless, cleaned up the parts crib, and introduced basic technical training."[10]

A plant supervisor at South Mack argued that the first plant manager could not assert social leadership either. "He oscillated between being participative and being a little Caesar. He did not know how to lead in a plant based on the team concept."[11] Nor were team members trained in the social skills needed to confront and solve conflicts. "Members were told to be like family communal groups."[12] Walton notes as well that the "team concept had come to be understood by team members to mean 'getting along' with other members . . . whereas originally it was intended to emphasize the interdependence of members' tasks, joint problem solving, and mutual help."[13] The supervisor argues that management gave the teams "all the difficult problems to solve, like discipline, but not the skills and assistance. An employee would be late, workers would complain, and we told them to deal with it themselves. Team members would come back and report that the late employee said, 'Screw you'; but we would say, 'You work on it.' " He notes that a critical incident in his own career as a supervisor occurred when he got one team to "think that they were *work* teams, not family groups. I made a list for them of what groups were. I told them that they don't have to like each other. I introduced task-focused skills to them, like meeting skills, brainstorming, force-field; skills that would help them get the job done."[14]

Management at South Mack did not provide the necessary technical and social leadership in the first year of the plant's operation. The level of developmental tension was too high relative to worker skills and plant structure. Consequently, when management introduced specialized jobs and increased the supervisor-to-worker ratio, the plant sociotechnical system lost some of its flexibility, its ability to provide learning opportunities. Management's response to worker difficulties was too little, too late. To get the plant on track, the new plant manager had to alter the job system substantially.

At Sound, Inc., self-managing teams were introduced in parts of the factory, but, as at South Mack, work group members were provided with little training or support. Supervisors and support personnel avoided involvement with the groups, "leaving them to a great extent to their own devices."[15] Supervisors feared upsetting the experiment in self-management by exercising undue influence, and few of them knew how to interact with a leaderless team. "If a supervisor was concerned that a worker was taking too many breaks, he had to tell the person directly, rather than pass the warning through the lead [the assistant foreman], who could discuss the situation with the worker more informally."[16] Similarly, when a manufacturing engineer "had a suggestion for a new assembly fixture, he either had to call a group meeting or tell one person on the line and rely on that person to convey the message accurately."[17] The team faced great production and quality-control pressure. Lacking management support, it created its own supervisory structure; as a result worker commitment to the team and the team process dropped dramatically.

The group had a rotating coordinator who scheduled materials and kept production records. Facing internal conflicts as well as production pressure, group members decided to expand the coordinator's role to resemble that of the traditional lead. Afterward, however, they refused to follow the orders of the first temporary lead, and several workers withdrew their names from the rotation list. They then elected a permanent lead, but discussion flagged at the next two team meetings. When a researcher-facilitator proposed that the group take up long-range strategic as opposed to operational issues, no one responded. "The silence was finally broken by a woman who had been extremely active in early meetings, but who had been very upset by conflict within the group. She was quite concise: 'We need a lead and we elected Lydia. Let her do it.' "[18] It seems likely that because management did not support the teams—because managers abdicated with the excuse that they were protecting the experiment—the workers ultimately abdicated their own right to self-management.

I suggest that management's decision to install an assistant coordinator at Big Chem was appropriate. The technical crisis was severe, the teams needed direction, and the coordinators needed support. The plant facilitator reports that some two years after the winter crisis the workers and managers remain committed to learning and participation. "We are doing a self-assessment, and large numbers of workers have participated actively in the review of potential assessment instruments and questionnaires."[19] He feels that workers have learned to acknowledge differences in competence among team members without engaging in conflict.

Clearly it is not easy for managers and workers to keep the plant at its optimal level of developmental tension. What seems like restrained leadership in one setting may prove to be management abdication in another. The warehouse manager of Trucking, Inc., expressed the ambiguity of leadership in such settings: "The main thing is, don't back off too much. We maybe backed off too fast. There is a long learning stage in everything. There were critical times where everything was going smoothly and then all of a sudden we were back down into a valley and the team managers had to jump back and change our styles again. You can't back off too much."[20]

More longitudinal studies are required to assess the ultimate impacts of management actions; for the present I propose the following tentative explanations for the failures at the plants under discussion. At Fall Mills excessive structure was imposed too early, reducing the opportunities for long-term learning. Management at South Mack and Sound, Inc., on the other hand, developed too little structure and did not support appropriate skill training. General Foods experienced a third failure mode: lack of challenge and novelty led to a decline in commitment and interest; plant management did not attend to the long-term consequences of operational stability. In contrast to all of these, the managers at Big Chem successfully handled a period of crisis by developing an appropriate level of structure.

How Workers Respond

Workers also shape the plant community's response to novelty and uncertainty. In most cases workers do not initiate broad programmatic responses (such as training programs), owing perhaps to the absence of unions in most sociotechnical settings. Rather, they respond to and in turn affect the level of developmental tension through more informal group processes. Two such processes are of particular importance: the creation of artificial divisions within and between work groups, on the one hand; and, on the other, the avoidance of necessary conflicts among workers and work teams.

Workers are supposed to evaluate one another's social and technical competence for pay raises. Often, however, work teams do not have the social discipline or cohesion to promote one worker and not another. At Fall Mills a worker would create his own subcommittee to evaluate his skills. The plant facilitator notes that "there was a lot of watering down, a lot of backscratching; the thing was breaking down. Teams liked making the good decisions but avoided the tough ones."[21] An internal consultant who helped design the plant believes that "workers simply could not keep money from another's pocket. What happened was a foot race. Individuals rushed to get passed on all the skills. People were just cranking out the minimum. They became 'jack of all trades and master of none.' We wound up with the highest box costs and the lowest production volume in the company."[22] At Big Chem one manager argued that "people were moving up too fast in the first year. They were not expert enough. Every eight months workers could go up one step. People would learn fast, but the downside was that people were focusing ahead and not on their immediate situation, like training others."[23] Finally, at General Foods one worker noted that "the match in the gasoline is pay." Another explained "that the tenuous basis for his security made him continuously concerned about his relations with the many people who could help him or hurt him in the future."[24]

Such tensions and problems are most often resolved by giving the supervisor or the coordinator formal or informal power in the evaluation process. For example, at Fall Mills workers evaluate one another, but the supervisor makes the final decision. "The team can still exercise a lot of pressure to get someone passed, but the supervisor has the last say."[25] At Family Food the supervisor and the personnel manager have the final say as to "who goes up or does not go up to the next level."[26]

The problem of peer evaluation is only one aspect of a broader issue. Can teams manage their own conflicts? If they cannot, if they avoid conflicts to protect group cohesion, then management may ultimately have to step in and limit the scope of the team's autonomy.

The problem of conflict management can undermine even limited sociotechnical designs. At City Bureau the city government and the union together introduced a quality-of-work-life (QWL) program.[27] A committee was established to review proposed changes in work design that could improve both productivity and the quality of work life. In contrast to full sociotechnical programs, the QWL program was not predicated on any particular concept of work redesign. Rather, it simply provided a forum within which redesign suggestions could be discussed.

At one committee meeting management suggested that sewer workers change their seven o'clock-to-four o'clock daily schedule to an eight-to-five schedule. The older schedule was a relic of the days before air conditioning. Most sewer workers favored the plan; but three, who had extra jobs that began at four o'clock, opposed it. Group members were unwilling to fight with the dissenters. At the next meeting of the QWL committee the sewer group announced that because it opposed management's suggestion it was withdrawing permanently from the committee. The worker representative told the management representatives, "If you was to make a plan giving the men every other day off for a holiday, they'd tell you to shove it."[28]

The workers' exaggerated response to a management suggestion was determined in part by the sewer group's inability to manage its own conflicts. The schedule change proposal created a difference within the group: some workers liked the idea, others didn't. But this difference so threatened group cohesion that the workers chose instead to withdraw from the committee itself.

This commitment to group cohesion can sometimes limit the learning and development of work team members. At South Mack the shipping and receiving team was having trouble controlling the inventory of metal parts. The shift manager asked the team to solve the problem. One member, a woman with a tenth-grade education, contributed the best ideas. "She set up a system for reordering. People came from our machining plant in X to see how it worked. But jealous team members told her she was toadying up to management and harassed her until she resigned." The South Mack supervisor believes that work groups sometimes overstress conformity and "group-think" and don't encourage "really way-out thinking."[29]

Finally, when team members are reluctant to recognize and address interpersonal conflicts, they can stalemate the team process. A supervisor at Fall Mills describes a situation in which a team was not working effectively. Its meetings were unproductive, and quality and production goals were not being met. "When I spoke to individual team members I was met with a smoke screen. They gave me the silent treatment. They wouldn't tell me what was going on." The supervisor went to a team meeting to discuss the problem but again received little response. His forays led him to a tentative diagnosis. Two members had become the informal leaders of the group but were actually hostile to the group setting and the group process. One had essential technical skills but was "a loner type, remote from the team. In conflict he would back down and sulk. He was aloof, distant." The other "was an authoritarian sort but was playing the role of the non-authoritarian." Anxious in the freedom of the work setting, he made use of his right to be open by constantly expressing anxiety and fear. "He was communicating his anxieties to the group about the fact that it was a team. The team couldn't confront them, couldn't make them work for the team, not against it. People started feeling incompetent. They didn't trust themselves or each other. That's why the team couldn't meet its goals." Ultimately the supervisor worked with plant management to remove the two from the team, and performance gradually improved.[30]

Creating Artificial Divisions
Workers sometimes cope with developmental tension and uncertainty by creating artificial differences between teams. For example, at Big Chem the Norm Review Board was composed of representatives from all the teams. In the first two years of the plant's functioning the board developed a handbook on good work practices and often reviewed complaints and grievances from the shop floor. It helped establish a body of "common law" to regulate behavior within and between teams. Yet early in the life of the board one team withdrew its representative. The team was known for its great technical competence, yet none of its members had succeeded in becoming assistant coordinators in other parts of the plant. Team members felt slighted and consequently left the board.

The facilitator to the plant argues that no team members became assistant coordinators because the team coordinators and a few key leaders dominated the group. They had great technical competence but were not letting others learn the more difficult control tasks. Consequently, team members could not develop a plantwide reputation. The team was felt to be too cohesive, too self-contained. The team was reluctant to accept rotations from other teams, and some said of it "that workers couldn't get in and couldn't get out." Tensions produced by an opportunity-poor group process within the team were ultimately displaced onto the rest of the plant, threatening plant cohesion and integrity. Because team members couldn't face their internal problems and conflicts, they fought instead with the rest of the plant.

Work teams often develop myths of their own superiority to lessen member anxiety about the many uncertainties that plant communities face. The superiority myth allows group members to believe that they will be immune from the consequences of any disasters befalling the plant. At Fall Mills one team developed an elite conception of itself. It was determined to become the first team to declare itself fully autonomous. It refused to accept transfers from other teams in order to protect its own development. After the team declared itself self-managing, the team manager was given a promotion, and team production quickly fell. A consultant from the parent company suggests that the manager's wish for promotion distorted the team process and that the group's feelings of superiority were covering up a dysfunctional group process.

Utopianism

The supervisor at South Mack argues that if teams are to prove successful, if they are to promote member learning and manage the machinery, they must be given tasks and responsibilities in stages. A certain utopianism infects both managers and workers in the first year of a sociotechnical design. At South Mack managers overestimated the work groups' abilities to handle difficult interpersonal issues. They were surprised when some workers continued to drink at lunch despite agreement among all members of the plant community that drinking could endanger those on the shop floor. "Most managers don't know how to handle the drinking problem of workers, and we are asking the teams to do it!"[31] He suggests that teams can effectively allocate their different members to different tasks after a year and a half of functioning, but that it takes four years for teams to make pay allocations between members. I suspect that management utopianism is a cover for management abdication. There is a hope that somehow the workers themselves will solve all the difficult problems and that managers will not have to define new roles for themselves in these new settings. Walton's observation that South Mack managers engaged in wishful thinking, believing that commitment alone could solve complicated social and technical problems, supports this hypothesis.

Worker utopianism is often expressed by the fantasy of rapid promotion for everyone, so that no one is disappointed. After six months at Trucking, Inc., according to the warehouse manager, "people were wondering why they were in the same jobs, doing the same work. There were a lot of promises made during the change period; or maybe expectations were conveyed by training personnel or . . . an academic approach [was taken] to it instead of a realistic approach. So within certain areas there was a bit of a downer: morale was going down." The manager argues for realism. "Don't promise them the world with this fantasy type of living where everything is peachy and rosy. By the end of the first year people might be expecting to take over the manager's position right away and might be wondering why they are not moving up within that organization to the main functions of the department."[32]

Utopianism is also punctured when workers face the difficult conflicts between individual and group needs. The sociotechnical system is designed to provide opportunities for individual development; yet its effectiveness depends on group behavior and performance. The rotation system plays a critical role here. Workers must often leave a team to give a chance to another person who wants to learn a new skill. The team, however, may want to keep its best technician to protect production and may therefore ask a less competent person to rotate onto another team. The worker who rotates may therefore feel demoted. At Fall Mills line workers who transferred to the maintenance team to learn mechanical skills felt punished and disappointed when they had to return to the line to make room for others. Out of spite some refused to fix broken machinery and called the maintenance crew instead. At Little Chem, which was moving toward a modified design with an explicit pay-for-learning system, management added two year-long coordinating positions for training and materials scheduling to one of the process teams. The two new coordinators "hoped it would be a permanent position. They were disappointed and sad when they transferred out to make room for others. They felt they were not good enough."[33]

Defining a Role for Management

Managers in sociotechnical settings, as well as workers, must develop new ways of behaving. Many managers are not sure how they should define and exercise their authority. They are, of course, ultimately accountable to the senior management of the company and must retain final control. Sociotechnical settings are not experiments in worker socialism or worker ownership, yet within this broad constraint they give rise to many ambiguous situations. If managers disagree with a team's decision, should they let the team make its own mistakes? If the team is reluctant to discipline latecomers, will other teams take this as a sign that tardiness is excusable? If, as at Trucking, Inc., some team members "are stronger than others" so that those others "take whatever jobs are left over," should managers intervene so that the strong ones stop taking the "cream-type jobs"?[34] If they intervene, will they be subverting the team's own autonomy, its right to regulate its own processes? Finally, if they face production pressures, should they scuttle the rotation system so that profits are protected?

Managers often respond to these ambiguities with guile and subterfuge, not because they are dishonest, but because they are confused. At Fall Mills the plant manager, worried about the performance of a team with not supervisor, did not communicate this directly and openly. Instead he told the team "moderator" (a team member who moderated group meetings) that "whichever team member has qualified in a production line skills group where there is a problem should help out . . . regardless of their present assignment."[35] This puzzled team members, since they were already helping one another out. Was the manager asking that they stop rotating? A few days later the manager sent a supervisor back to the team but called him a consultant. One worker responded, "he says he is a consultant, but I know he'll behave just like a manager."[36]

I believe the manager was afraid to be direct. He feared that if he asserted his authority directly and openly, the workers might interpret his decision as an attack on the sociotechnical design itself—as an attempt to overturn the new, though ambiguous, norms that governed the day-to-day relationships between workers and managers. By revealing his dissatisfaction in stages, by softening up the work group for his decision, he hoped to minimize the reaction. The workers, however, could see past the word "consultant." When managers engage in such tactics, they may paradoxically create more worker mistrust just as they are attempting to protect the overall sociotechnical design.

Manufacturing engineers at Sound, Inc., experienced the same confusion. The engineering supervisor observed, "My engineers are having trouble relating to this group [the autonomous team]. They are afraid to say anything, afraid to do anything that is going to upset them. They're afraid to say something that is going to screw up the whole works. Everybody is a little nervous. I'm reacting myself, you know, because I like to kid a lot. I'm almost afraid to kid with those people, because they are already in an upset condition and I might really trigger some animosity."[37]

The same problems of communication occur when management rejects a team decision indirectly. " 'Noes' begin with the phrase, 'but are you sure you have considered these additional factors?' " Workers "interpreted these words as being management's way of saying 'no' under conditions of self-management."[38] As one worker at Fall Mills put it, "Management hardly ever says 'no' right out; but when they ask us to rethink our decision because of 'other factors,' we know they want another decision."[39]

These communication problems between managers and workers are mirrored by communication problems between senior plant managers and supervisors. In sociotechnical settings supervisors, often called facilitators or coordinators, are supposed to help the teams develop and learn. They are not supposed to control or command them. Yet often supervisors feel that they are nonetheless ultimately responsible for the team's performance. At Reading Meat the cutting teams had to respond to rapid changes in meat and poultry prices due to sales by competitors. When a meat sale was planned for the coming week, the team was supposed to figure out what to cut, what to vacuum pack, how to divide up the work, and how long the cutting would take (vacuum packing the portions of the carcass not on sale considerably lengthens production time). Faced with a deadline, the staff assistants were unwilling to give the planning task to the group. Rather, they figured out the work schedule themselves, discussed it with the work assignment and cost coordinators, and then instructed the team. Similarly, the staff coordinators kept the most skilled cutters in the "bull-saw" position, where the first cut was taken, and prevented others from rotating through it. The first cut could determine the market value of the whole carcass; in a competitive market with fluctuating prices for beef and chicken, the coordinators did not want to take any chances by rotating workers. Thus, because of time and cost pressures, many group members did not develop planning or cutting skills. Senior management did not step in to change the supervisors' behavior.

The facilitator at Reading Meat convinced senior management to change the staff coordinators' assignments. He argued that they should not be in charge of particular teams but rather in charge of particular functional areas, such as cost control, supplies, and training. They would work with committees composed of team members who had corresponding duties on their own work teams. This would have three effects, he suggested. First, the coordinators would not feel directly responsible for the teams' performance and therefore would not take over the teams' decision making. Second, the coordinators would take their functional committees seriously. Third, work group members would take seriously the specialist positions within the team, such as training, cost control, and work scheduling.

At the time of my interview the facilitator reported initial positive results. "Earlier many of the specialist roles within the teams were just not functioning. Elections to them tended to be more like beauty contests. But now team members are taking them more seriously."[40]

Such "functional" supervision designs can bring problems of their own. At Fuel Inc., each supervisor was responsible for a particular function (such as supplies or health and safety) rather than for a particular team. Unlike those at Reading Meat, they were keeping their hands off the teams. The plant manager was not happy, however: he believed that the supervisors were not monitoring the teams. Lacking information about them, he didn't know how the teams were working. He felt that the supervisors were "too comfortable," and that made him nervous. The supervisors responded that the teams were performing well and that production and quality were up to specifications, but the plant manager wasn't satisfied. A facilitator suggested that the plant manager felt that "the supervisors were not worried enough."[41] Because traditional foremen are very active, solving technical problems, ordering supplies, disciplining workers, and monitoring the work flow, senior managers are made uneasy by the low visibility of supervisors in sociotechnical settings.

Supervisors are aware of this. At Fall Mills one supervisor noted that his most anxious moment in the plant's first year occurred when his work group took an overlong lunch hour. The incident was trivial, but other supervisors noticed it. In fact, his group, with an advisor from another plant in the parent company, was celebrating its excellent work on a technical problem. The supervisor felt that however trivial the infraction, management would learn of the breach of rules, his credibility would fall, and he would have to work to reassure upper management. He went to the team and told them that "violating a clear guideline was not acceptable."[42]

Senior managers may not appreciate the complexity of the relationship between a supervisor and a team. They may expect that supervisors can easily take over when necessary, though supervisors are frequently reluctant to interfere with the team process. As one manager at Sound, Inc., said of a supervisor, "He's still responsible to make the production schedule. I'm not coming down on anybody but [the supervisor] when the schedule is not met. This morning in that brief meeting we had . . . he made the comment, 'If it's okay [with the team], I will do this.' And I thought of myself, 'What the hell you asking, if it's okay? You're responsible for that area.' "⁴³

Finally, supervisors find it hard to balance the competing demands of team performance and team learning, if the teams themselves do not accept responsibility for their affairs. As we have seen, teams will refuse to regulate themselves if this autonomy produces too many conflicts and threatens group life. But workers will also resist self-management and its attendant risks if they feel that senior management is not fully committed to the design. Two supervisors at Office, Inc., were surprised that workers did not want to participate in a QWL program. "They were sitting back, they weren't coming up with anything. There was a time when the managers and supervisors that believed in the project had to push them. The employees didn't seem to want to come out with anything because of that feeling of lack of trust."⁴⁴

The level of trust at a given plant is determined in part by its history of labor-management relations. Where these have been hostile, workers will see QWL programs as management gimmicks and will resist such programs, suspecting that management isn't serious—that if the going gets rough, management will pull out. At Office, Inc., where labor relations were actually quite good, junior and senior managers proceeded with great caution when introducing or proposing any new program. "We always listed in our proposed projects the advantages and disadvantages, covering all eventualities before we started."⁴⁵ It was this caution, I suspect, that limited worker responsiveness: if management was so cautious, why should the workers take the risk of trying something new? Participation is not a gift from management to workers. On the contrary, it can come at great cost to workers, who must risk interpersonal conflict. Workers must evaluate their peers, determine their pay, and discipline them. They will not begin such an undertaking if they believe that management is not fully committed to the new design as well. In any work redesign workers and managers come to a point of no return. They must either decisively depart from previous organizational practice or accept a very narrow conception of work reorganization.

Sociotechnical designs can founder when the interconnected dysfunctions of teams, supervisors, and senior management create a vicious circle of failure. Imagine the following hypothetical situation. Senior management, doubting the teams' competence, removes certain crucial quality-control and maintenance jobs from the rotation system. Workers, perceiving that senior management doesn't trust them, become less willing to take the risk of interpersonal conflict and lower their standards for evaluating one another's skills. After a machine accident senior management realizes that workers are not as skilled as the evaluations suggest. Supervisors are ordered to oversee the evaluation process. The workers then cry foul. On one team an informal leader asks the supervisor to give the team another chance. The supervisor, hoping to save the design, agrees. When senior management hears of the supervisor's behavior, it orders him to intervene. He steps in, and the informal leader accuses him of betraying their agreement.

Support Groups

The relationship between managers and workers will be mediated by the behavior of support groups such as buyers, salespeople, engineers, and accountants. They are rarely included in the team system and are frequently hostile to the sociotechnical experiment. Because they can no longer go through a visible chain of command to supply teams with resources or to obtain the information and compliance they require, their work becomes more complicated.

When two shifts at Sound, Inc., cooperated to resolve intershift problems in the lacquer finishing area of the plant, engineering personnel and plant superintendents, who viewed participation as "socialism," blocked the proposed solutions. Even members of the support staff who wanted to cooperate with the teams found it difficult to do so. A quality-assurance supervisor posted the results of her test without talking to anyone on the team. She didn't know who was in charge and could not effectively work with a leaderless group.[46] At Household, Inc., managers in various parts of the company started a number of experiments over a ten-year period, many of which failed owing to lack of support from senior management. In a retrospective survey, workers who had participated in the experiments said they felt support staff should have been invited to team meetings.[47] Accountants at Scott Compressor found the team system an added burden: the financial data base for the production system was organized to produce team-based information, which was hard to integrate with the accounting requirements of the parent corporation.

Conflicts between the teams and support units may arise when the latter fail to understand or accept the autonomy and authority of the former. The warehouse team at Reading Meat had the authority to reject damaged meat. In one shipment the skid on which the carcasses were placed had fallen over in the truck. A worker told the buyer that he could not accept the meat; the buyer, anxious that the stores meet their commitments to a sale the following week, pulled rank, asking the worker if he was "prepared to accept full responsibility for rejecting the shipment." The worker backed down; he told his team that the team work design was "bull."[48] Often support staff cannot adjust to working directly with an operator. They insist on "seeing the boss," although none exists. This can anger and demoralize the work team.

In a similar case, buyers at Trucking, Inc., were angry when a team, having finished assembling a load of shipped products, knocked off early one day without unloading more trucks. But because the team was planning its own work and setting its own goals (consistent with efficiency), the workers were free to leave when they had completed the day's tasks. "The buyers just don't understand that if I didn't let them go home early, the assembly would have taken the whole night."[49]

Finally, the relationship between people in the team system and support staff is often affected by the relationship between the socio-technical plant and the parent company. At Smart Chem, for example, managers and internal consultants tried to get a higher salary for the plant supervisors, but the personnel department at corporate head-quarters would not agree. There was no distinct job description in the department's files for a supervisor in a sociotechnical setting. Weren't they just like foremen? The plant manager argued that they had a much more complicated task, but the personnel office did not budge and senior management did not intervene.

The response of support groups ultimately reflects senior manage-ment's attitude toward the sociotechnical design. At Sound, Inc., support group resistance was a sign of senior management's tentative com-mitment to the work team experiments. Managers did not use their influence and authority to force support units to cooperate more ef-fectively with the teams. A plant manager who has full control over all functions within the plant (production, maintenance, quality as-surance, purchasing, accounting, and so forth) and who is committed to the design can ensure that individuals not included in the team concept nonetheless cooperate with it. But such a plant manager will get only lukewarm support from many senior managers in the parent company who resent the new ideas and methods that the experiment brings to the company. These senior managers may not be powerful enough to close down the experiment but are powerful enough to exert pressure on the plant, monitoring it closely, watching for any signs of failure.

Careers

Team conflicts, support unit hostility, and the relation of plant management to supervisors all help determine the short-run level of developmental tension in a plant. In the long run, however, the issue of career mobility for workers will prove decisive. Developmental tension will fall if workers have mastered all the skills. When most workers earn the top rate, might not learning and commitment fall as well? At General Foods "there was a long period, after the dry plant reached an operational steady state, when the absence of externally stimulated changes, no new products, no new expansion, no new process technology helped perpetuate . . . a sense of stagnation."[50]

Two options are available at present. First, managers can include in the pay-for-learning system skills and jobs that take workers off the shop floor. When they master the blue-collar skills, they can go on to learn the white-collar skills and so advance in pay and grade. For example, at Trucking, Inc., the job of the traffic clerk, usually not performed by warehouse manual workers, was moved into the team rotation. Similarly, at Scott Compressor work team members now rotate into clerical positions in small offices placed beside the major work stations.

Second, it is often possible for management to respond to the surplus of worker skill by introducing new products that can increase the plant's profitability and broaden its market. These new products may in turn require new control and maintenance skills. Workers can be rewarded with increases in pay and grade. When many workers become multiskilled, the technical adaptability of the plant is high. Thus Farm, Inc., reached a technical steady state about three years after it started operating. Although most workers were not at the top rate, the work teams and management had solved the critical social and technical problems. Worker talents and creativity could be deployed to new tasks. During an unforeseen slump in the market for grain dryers, management decided to produce fans in order to avoid a layoff. "The manufacturing of this new product started rapidly and with great effectiveness," wrote one analyst. "This enhanced systems-response capability is especially beneficial to Farm, Inc. Their business, primarily in agricultural products, is cyclical, often with expected fluctuations. The capacity to continue productively through changing demands is therefore especially desirable."[51] The very adaptability of a sociotechnical system may become the vehicle for expanding the skill and knowledge base of its members.

Yet in the long run it is clear that no setting—no sociotechnical plant, conventional factory, or corporate office—can provide continued growth for everyone. In sociotechnical settings all workers can reach the top rate; there is no contrived scarcity of positions such as exists in traditional hierarchies. Nonetheless, those workers who have reached the top rate and wish to continue to grow and develop must look for opportunities outside the plant. Often they leave sociotechnical settings to get jobs as plant managers in other companies. Skilled workers from Fall Mills, for example, got management jobs in neighboring plants. But sociotechnical factories might design more explicit "graduation" processes into the plant system. The manager at Family Food has worked with the senior management of the parent company to place his graduates with other companies in the parent system. A sociotechnical setting functions partly as an educational institution, actively promoting the learning and development of its members; the next step would be for it to develop career guidance and placement services. By linking up with factories in its region, developing norms of graduation ("most people leave this place after seven years"), and providing guidance and placement assistance to its workers, it might continue to promote the careers of its older workers while sustaining an atmosphere of learning and commitment for its newer members. I know of no plant that presently does this; but as the number of sociotechnical plants grows, and as they develop mechanisms for linking up with each other, such innovations will take form.

In Sum

We have seen the various ways in which sociotechnical systems fail. "Failure" should be understood here as a relative term: none of the firms I studied has failed to be profitable over the long run, nor has any plant (as of the date of this writing) regressed to a conventional chain-of-command system. Rather, these examples show the ways in which plants respond to excessive developmental tension by creating rules, procedures, and regulations that limit the potential flexibility of the plant systems and constrain their initial open-ended framework. Moreover, it is not surprising that such plants frequently fail to attain an optimal level of developmental tension. They are social innovations and operate under considerable disadvantage: weak support from the centers of corporate power, a lack of tradition and history, and the need to develop novel practices and cultural norms. Like innovations in many fields, the early designs or models frequently fail.

Five modes of failure have emerged as significant.

1. Management responds to developmental tension by reducing team autonomy. This limits the range of skills that team members can acquire and ultimately reduces the flexibility of the plant system.

2. Management abdicates its responsibility to the plant and the teams. It does not support the teams with either training or temporary rules that might enable team members to continue to learn without damaging the technical system or reducing profitability.

3. Relationships between plant management and supervisors are poorly articulated. Too often supervisors feel that they are overtly asked to support team development but covertly encouraged to take over the team at the first sign of difficulty.

4. Support units undermine the team system: either they are hostile to it or they don't know how to operate within it.

5. Workers respond to developmental tension in one of two ways: Either they ignore necessary differences between teams or team members, or they create artificial (non-task-related) differences between teams or members.

Lessons for the Future

As new and old plants develop learning systems based on sociotechnical designs, they can benefit from the experiences of the pioneer plants established in the last decade. Managers in particular must address the complex process of allowing teams to develop and become autonomous while providing them with the training and temporary regulations that support team development. Two issues are particularly likely to cause trouble in the early years of a plant's development.

First, peer evaluation can greatly strain the team's social processes. Workers should participate in promotion and pay decisions, but managers should have the final say. If a team feels that the supervisor's decisions are consistently unfair, it should be able to take its grievance to a plantwide forum, such as Big Chem's Norm Review Board.

Second, managers should be alert to pressures from either workers or supervisors to split the skill areas into distinct career lines. It is likely that most plants will develop a system where workers major in one skill cluster and minor in others. But the plant community reduces its flexibility and coherence when it completely separates these clusters.

For example, if nuclear power plants separated one skill cluster from another—particularly maintenance from operation—they might replicate the problems of conventionally designed plants. In utility companies the quality-assurance group is most often separated from the line organization, safety procedures are reviewed by inspectors who are not part of the production team, and craft boundaries separate maintenance workers from process-control workers. Yet we have seen how important it is for workers in such plants to develop an integrated understanding of maintenance, safety, and operational tasks. If workers can rotate through all these tasks, the flow of information from one function to another will be secured.

What Is Distinctive about Sociotechnical Plants?

Settings that integrate work and learning are culturally distinct from traditional factory settings; relationships between people, and their perceptions of these relationships, are altered. Three characteristics of the new systems are critical.

First, the sociotechnical system intensifies the political and interpersonal dimensions of group activity. Beth Atkinson has studied the politics of team systems by videotaping and coding work-team meetings at a fuel plant.[52] She sought to understand how leadership and authority are exercised in teams when positional or formal authority is weak or absent.

Atkinson found that a team member could exercise leadership by drawing strength from a particular power base within the team. Bases included such skills, attributes, or resources as information, expertise, connections, energy, creativity, and charisma. Workers exerted their power through both supportive and abusive tactics. For example, workers with an information base could give or withhold information, workers with an energy base could demonstrate enthusiasm or stubbornness, and workers with a communications base could clarify or put down other people's comments.

By coding verbal transactions during a meeting, Atkinson could measure the power balance in a team, the degree to which power was or was not distributed evenly among team members. In certain teams a few members dominated the group, frequently using such abusive behavior as cutting off comments, ridiculing suggestions, and blocking conflicts. Such members often developed close alliances with a few others on the team, who then protected the leader's dominating position.

People who use charisma as a base produce the most stifling group climate. Atkinson notes, "This is the deadliest power base. The person rules the group by exhibiting abusive behavior. People want to be liked by the person. . . . It is a hard group to crack and it creates difficulty for the members and for people who want to change the group."[53] In groups with a narrow distribution of power the powerless were quickly and permanently stereotyped.

Atkinson believes that groups with a broader distribution of power are more effective. In one experiment she tried to change the power balance of several groups by teaching them to recognize and point out abusive behavior.[54] She gave each team a brief lecture on power bases and then asked members to flag abusive behavior (actually by raising a small red flag) during the team meeting. After the meeting she presented her own count of abusive behaviors. Her count was always greater than the team's count, but team members always agreed that hers was the accurate one. Some teams responded well to her intervention. They worked to reduce abusive behavior in subsequent meetings and so redistribute power in the group.[55] These teams "were recognized by others in the plant as having improved their team functioning."[56]

Power theory supports Atkinson's finding that power distribution and team effectiveness are correlated. Both relate critically to processes of disclosure and confrontation. A "component related to effective team functioning is willingness to risk disclosure of ideas and feelings. Shutte demonstrated that equal power resulted in maximum disclosure. Unequal power, on the other hand, affects both the high-power and the low-power person in a negative fashion."[57]

In turn, Argyris argues that disclosure and conflict are intimately related.[58] People avoid disclosing ideas and feelings precisely because they fear that the group will prove unable to handle the resulting conflicts. Thus they fear the consequences of their own presumed aggression. As we have seen, team members in Atkinson's experiment consistently underestimated the level of abusive behavior in a team. "The intervention that members used in abuse reduction [flagging the abusive member] required both cooperation and courage."[59]

Team members will be better able to confront one another if individuals feel that they have a power base, if they have been able to exercise authority and leadership in the past. Teams with a narrow distribution of power will produce minimal confrontation and disclosure. Often, however, team members must confront and disclose if the team is to develop. The workers in one team at Fall Mills, for example, could not confront two informal leaders who were blocking the team's functioning. One had a base of expertise and used the abusive behavior of withdrawal; the other had a base of communication and used the abusive behavior of giving long speeches. The supervisor, who had both positional authority and a base of expertise, had to step in and solve the problem. Thus team functioning, group development, the willingness to confront others, and the distribution of power are all closely correlated. A team's effectiveness rests on the political and interpersonal skills of its members.

To be sure, every work group produces a micropolitics of sorts; but while in traditional settings group processes are checked by formal authority, in sociotechnical settings teams rely on these processes to solve problems and promote member learning. Team members must develop a deeper awareness of these processes. They must become aware of how they take or yield leadership in unstructured settings, how they confront or refuse to confront others, and how they succeed or fail in disclosing group issues.

A second characteristic of sociotechnical systems is that they simultaneously emphasize the principles of both individual and group performance, both individual and group reward. On the one hand, the team is the unit of performance and individuals learn to regulate their work in the context of group requirements; as we have seen, individuals become sensitive to group processes. On the other hand, each individual is encouraged to develop, to learn, and to create a career that may lead beyond the team and perhaps beyond the plant. Just as individuals must accommodate themselves to the group, so must they develop a concept of themselves as people with distinct careers.

In traditional manufacturing settings the worker is alienated from the group at work. His personal accountability to the group is limited by the presence of supervisors, who ensure that his work is consistent with the work of others. At the same time he is alienated from his own capacity to learn and develop. He has no career. If he wishes to develop, he has to enter a craft, which automatically restricts the scope of his learning. Thus in the traditional factory neither group needs nor individual needs are emphasized.

Two other alternatives are represented by other work settings. The current managerial career promotes individual learning and development (fast-track managers are exposed to many different divisions of the company) but not commitment to a specific group or team. In contrast, a disciplined army unit or the Japanese work system promotes group loyalty without emphasizing members' individual development. Many analysts point to the Japanese model of industrial relations to argue that our work culture must increasingly be based on the principle of accountability to the group.

The sociotechnical system achieves a new and unique integration of individual and group needs. It does not balance the two; rather, it intensifies and highlights both. This is why such settings are particularly stressful, but also why they are flexible.

Third, sociotechnical settings function as schools where teaching and managing are integrated. Not only do workers learn new skills, but managers, supervisors, and workers develop a conception of themselves as teachers. At Smart Chem the supervisor felt that he had to develop a theory of how other people learn in order to guide his own behavior. Working with an external consultant, he devised a "level of thinking" typology to guide his dealings with workers.

There is a hierarchy of questioning. A worker can ask me the "how" of something, the "what" of it, or the "why" of it. . . . If you ask me *what* an internal combustion engine is, I can describe its parts and function; if you ask me *how* it works, I can describe its internal dynamics; if you ask me *why* we are using it, I have to describe the production system of which it is a part.[60]

The supervisor believed that unless he was sensitive to the level of the question asked, he would either overmanage or undermanage the group. He linked his management behavior to his role as a teacher and to a theory of how people learn.

Just as learning opportunities are developmental for the worker, so teaching opportunities are developmental for supervisors. At Scott Compressor a manfacturing engineer was reluctant to cooperate with the teams, to engage them in problem solving.

When he worked with the maintenance men, he didn't do it from a developmental point of view. He couldn't teach. A lot of the workers resented it. There was a lot of conflict around this. He would say, "I am fifty-five years old, I came up through the ranks, I'll do it my way." The managers used him as a resource to develop a training program for the operators, and that began to change his mind. He became interested in how you teach people. Later, when we were redesigning the salary system for the nonexempts, he was called upon as an expert witness. He was clearly making statements in favor of learning criteria rather than just performance. He appreciated learning as a process. This was a real sign of change.[61]

I expect that this schooling dimension of sociotechnical settings will be recognized increasingly by other institutions and organizations. The federal government is interested in retraining for the new technologies. High schools are trying to develop vocational programs that will take poor teenagers beyond the knowledge of obsolete craft skills (such as printing). Both may turn to the sociotechnical factory for education and retraining. These factories may in turn attract educational subsidies that will allow them to expand their teaching functions without compromising production and quality.

Coordination without Hierarchy

When machines are linked together to form manufacturing systems, automatic controls regulate the movement of materials between individual machines. Machine integration can create system failures, since each subunit must present the material or product in process in just the right manner, with just the right specifications, to the next machine. There is no time for manual adjustment. Automatic controls adjust the work in progress so that such failures do not occur. First-order errors, the errors that emerge with machine integration, are limited by first-order control systems that import error into the feedback loops themselves, so that random variations in ambient conditions or product characteristics can be corrected before they distort or deform the final product.

But these integrated machine systems create new modes of failure, or second-order errors. Control-system failures and the new stresses placed upon these machine systems create both unexpected breakdowns and planned discontinuities. The work group must become the new second-order control system. Work teams must import error into their culture of work; they must learn and learn to learn.

Yet postindustrial technologies may seem to pose insuperable obstacles to the design and functioning of such work groups. Centralized "command systems" seem most appropriate when dealing with technologies that fail, often suddenly, in unexpected ways; groups who manage high-risk systems, such as submarine crews, missile-base platoons, and prison guards, are traditionally organized hierarchically. On the other hand, these same technologies enable the factory to respond flexibly to changing market conditions. In order to make full use of this capacity, workers must take initiative, learn, and share their knowledge and information with other units and divisions. Workers will most likely take initiative and responsibility when authority and responsibility are decentralized. How can we design organizations that are both centralized and decentralized?

The sociotechnical design is one example of an organizational form that transcends this dichotomy. It permits coordination without hierarchy. It does this by locating initiative in the teams, by providing cross-talk and rotation between the teams, and by supporting a plant governance system that factory members use to monitor the team system and its relationship to the factory's environment. This design can work only if it is embedded in a factory culture that supports learning and development.

13

Can It Happen?

The new technology and the new experimental factories together give a coherent picture of a postindustrial work system. Together they show how social and technical systems, when appropriately combined, integrate learning and working. Such systems enable workers and managers to respond to unique events, opportunities, and failures.

The new settings are innovative responses to discontinuities in machine design, in worker-machine relationships, and in person-to-person relationships at work. At Big Chem, for example, workers controlled automated machinery, used innovative control panels, and worked without supervisors in small teams.

While the number of such settings has grown rapidly over the last decade, most factories still use old machines and traditional supervision systems. Our society cannot easily appropriate the new technologies: industrial structure, engineering culture, career dynamics, and corporate financial policy stand in the way. Each discontinuity is embedded in a more inclusive social system that limits the new technology's impact.

The Discontinuity in Machine Design

Mechanical machinery is designed to eliminate error and slippage. The systems of transformation, transmission, and control are tightly integrated to ensure coherent and economical machine systems. Electricity allowed engineers to separate the transmission system from the systems of transformation and control. Feedback mechanisms, embodied first in mechanical and pneumatic devices and later in electronic ones, enabled engineers to separate the control and transformation systems. These developments reduced the level of constraint within the machine and so created flexible but productive machine systems. Cybernetic machinery could adjust itself in response to changing ambient conditions. Microprocessors have extended this capacity for flexibility. Workers can adjust the controls of machine tools to produce unique patterns of movement. Feedback devices and computer software together create machines that can be used to produce a large number of products under a wide range of ambient conditions.

Metal-Forming Technologies

Cybernetic controls were first introduced into the petroleum-refining process; today the petrochemical industry is the most advanced sector of the economy. The new technologies are most easily applied to flow processes. The movement of liquids provides a natural basis for continuity, while controls need to be installed only at critical junctures (where a catalyst must be added, for example, or where liquid must be heated).

Creating a continuous-process metal-forming technology is much more difficult. Cybernetic systems have not been extensively applied to metal forming—to the production of cylinders, boilers, engines, valves, cutting tools, flywheels, and dies. The workpiece does not naturally flow, and it goes through many transformations as it is cut and shaped. Controls must be introduced to monitor the cutting tools as well as the workpiece. Finally, it is more difficult to model the dynamics of metal cutting metal than to model the flow of liquids.
The American machine-tool stock is aging. Two-thirds of the stock is more than ten years old, and half of that is more than twenty years old.[1] Forty percent of all manufacturing workers are in metal-working shops,[2] while only 5 percent of machine tools are computer controlled.[3] One author who studied a machine shop notes that the "agglomeration of speed drills, radial drills, vertical and horizontal mills, chuck and turret lathes, grinders, etc. could be found in essentially the same forms in machine shops at the end of the nineteenth century as they are today."[4]

Many machine shops cannot afford to invest in new equipment. In the United States there are twelve thousand die-casting businesses whose annual revenue averages only $500,000.[5] The price of a ladling and extracting robot—one that pours molten metal into the mold and extracts the finished part—is $35,000.[6] Since most shops have between five and ten ladling and extracting machines, the cost of replacing old machines with robots could be as high as 60 percent of total revenues. Moveover, "if the robots were to be installed in a typical factory, all the other equipment would have to be upgraded to accommodate them."[7] Finally, many machine shops sell most of their products to automobile makers. This limits the range of their products and reduces their incentive to invest in a more versatile production system.

The Machine-Shop Work Culture

Economics is not the only constraint. The work culture of machine shops is inconsistent with the technical parameters of continuous-process machinery. Many machine shops are organized on a modified piece-rate system: workers get a base rate and a bonus that depends on the number of pieces they produce. Work relations are organized around workers' attempts to "make out," that is, to get paid at a rate between the minimum and the allowable maximum.[8] To make out, workers must manage a complicated set of relationships with cribmen, truck drivers, inspectors, foremen, and workers on other shifts. The truck driver may not deliver materials to the worker on time; the inspector may delay inspecting the first piece of a run; the cribman may give the worker the wrong tool; the worker on one shift may not tell his counterpart on another some of the angles and tricks for working a difficult job.

Michael Burawoy, who studied and worked in a typical machine shop, calls making out a "game."[9] He argues that the game is more elaborate now than it was thirty years ago, because managers have retreated from the shop floor. Workers regulate their relationships with only minimal interference from foremen and industrial engineers. The methods engineer no longer appears on the shop floor looking for loose rates, the foreman is more cooperative and lets workers chisel so that they can make out, and senior workers can get jobs with looser rates. Workers negotiate less with managers and more amongst themselves. Management pressure has fallen.

Labor shortages and turnover have played a role here. In pre–World War II shops senior workers could not advance; there was no internal labor market. The job structure was undifferentiated and a job transfer meant that a worker had to learn to work a new machine. Workers quit rather than move to another machine. After the war, machine-shop owners differentiated the job structures so that better jobs had looser rates. Workers tolerated jobs with tight rates, hoping to get easier jobs as they got seniority. Consequently, turnover fell and owners did not have to bear the costs of recruiting and training new workers. This became especially important with the growing shortage of skilled machinists. Today skilled welders and tool and die makers are in particularly short supply.[10] Many owners complain that apprenticeship programs are inadequate, that they fail to attract talented and educated students. As one analyst writes, "It is becoming increasingly difficult to attract apprentices to the metal-working trades. Young people seem to prefer other types of jobs, particularly in the service sector of the economy. This is a worldwide development; many European countries are already forced to import large numbers of semiskilled and unskilled industrial workers."[11] Labor shortages and labor turnover have thus reduced management's presence on the shop floor.

The elaborate game of making out makes it difficult for workers and managers to use the new technology. Burawoy notes that a computer-controlled cylinder-block casting machine was out of order much of the time "despite or because of its sophistication. . . . In some departments one or two computer-controlled machines had been installed, but they too seemed to experience considerable downtime."[12] New machinery that requires new skills and new relationships between people undermines senior workers' ability to make out. In a piece-rate bonus system, workers do not solve problems in teams; individuals are responsible for making the machines work. Unless a worker can negotiate informal support from others, he may be penalized when working with a new machine. The new machines can thus upset the existing balance of personal coalitions and established rates. Burawoy tells the story of a worker who struggled and failed to work a new and complex pulley-balance machine that he was supposed to use to locate imbalances in pulleys.[13] "The machine . . . is very sensitive to any faults in the pulley, faults that other machinery operations may inadvertently introduce or that may have been embedded in the original casting when it came from the foundry."[14] All the workers had trouble balancing the heavy pulleys: they were hard to lift and often did not fit properly onto the fixture. Burawoy notes that he and his co-worker "tried to pretend that the pulleys weren't there," even though "there were always a good number sitting by the balance."[15] Burawoy's co-worker hated the machine. "It wasn't so much that the pulleys were not offering him enough money, since Bill would have his time covered with a double red card. It was more that he had been defeated, his job had been taken over, he had lost control."[16] The new machinery threatened Bill's ability to make out. Because he had failed to work the pulley balance, he might lose status on the shop floor and his relationships with support workers might be jeopardized; and if he lost the double red card (since after some point the machine would no longer be considered to be new), he might not make a sufficient bonus.

The Management Game
The game of making out limits management's ability to introduce new machines on the shop floor. Indeed, when managers try to make

changes they paradoxically create intramanagement conflict. A quality-control manager might, for example, announce that scrap must be reduced by 5 percent. This would mean, however, that inspectors must spend more time looking at the work and that workers must take the trouble to report the amount of scrap actually produced, both of which would upset the delicate balance of relationships and rates that organizes the game. Production rates would fall, and production management would put pressure on quality-control management to relent. The shop-floor game thus functions as a homeostatic device, absorbing or buffering management edicts and transforming them into intramanagement conflicts. The persistence of such interdepartmental conflicts makes it difficult for managers to coordinate their interests in order to gain control over the shop floor. Finally, interdepartmental conflict leads one department to impose rules on workers in order to demonstrate its power relative to another department. A failure to change shop-floor behavior through new rules creates more interdepartmental conflict. This chronic conflict is management's game and parallels the workers' game. Both games persist because the management system and the shop-floor system are only loosely coupled.

The game of making out does not depend on conscious teamwork. Rather, like a market, it evolves out of a set of relationships and tacit norms. Since the game is unacknowledged, its players cannot step out to discuss the rules or to adapt it to technical discontinuities. The game does not contain rules for creating new rules.

Profits, the Work Culture, and the New Technology
Machine-shop technology is hard to change. Shops produce many of their parts on short notice, in small batches, and for local markets. Competition from large national and international suppliers is unlikely to drive prices down. As long as owners make adequate profits, they have little incentive to invest in new technologies. Moreover, if they install new machines, they must transform the relationships between managers and workers and among the workers themselves.

The new technology is fundamentally inconsistent with a work culture based on piece rates. Cybernetic technology involves continuous processes by which machine tools are either directly linked or indirectly coupled through a moving pallet. In such a process workers do not work on a single machine or job. Instead, they control a bank of machines and become process operators. Piece rates have no meaning, since no single worker is responsible for a particular piece. The whole game of making out becomes obsolete. Instead of competing with one another, workers are drawn together in process teams and have the opportunity to construct their relationships with one another more self-consciously.

The old technologies are perpetuated in a vicious circle. Traditionally, a labor shortage leads owners to invest in new machines. But in this case the labor shortage has led managers to abandon the shop floor to the game of making out, and this game limits both managers' and workers' ability to use the new technologies. Yet one positive element must be considered. The elaborate game of making out is a sign of workers' latent abilities to manage complicated interpersonal relationships at work. As management has withdrawn from the shop floor and given workers greater autonomy, workers have constructed an interpersonal game of great complexity and robustness. Yet their skills are bound up in an obsolete technical apparatus and a primitive remuneration system. It is as if a latent postindustrial work culture, indicated by the workers' interpersonal skills, is both embedded in, and perverted by, this obsolete technical apparatus. If the logjam can be broken, the new technology may thrive in an already present, if currently deformed, postindustrial work culture.

The Discontinuity in Worker-Machine Relationships

Automatic controls and Taylorist job design together create the concept of self-regulating machine systems. Engineers assume that such systems can function under a wide range of unpredictable environmental conditions without human intelligence to guide them. Taylor and his disciples eliminated the craftsman's skills by rationalizing his work. The automatic controls displace the worker from the production process entirely—or so it seems.

This notion of total machine control is false. As we have seen, engineers have not attained the necessary technical and mathematical understanding of complex production processes to develop comprehensive control systems—systems that can keep cost, volume, and quality at optimum levels. The production processes are hard to model, many relationships between crucial variables are only intuitively understood, and engineers cannot predict how extreme values of certain ambient conditions (such as the quality of the raw materials) might change variables and relationships within the production process. The vision of total machine control gives way to a concept of learning in which workers' tacit knowledge and engineering expertise combine with data tracking by minicomputers to progressively enhance the control system.

Indeed, it has become apparent that the control systems themselves create new sources of failure and error. As software systems that regulate the controls are modified and improved, and as people learn to exploit their potential, new errors come to light. Hidden faults in the software design become visible, while continuous modification makes the software system increasingly rigid. In addition, the flexibility and integrity of the machine systems themselves engender new patterns of failure. Small mechanical or electrical failures can ramify throughout the machine system. New patterns of interaction between machine-system parts emerge that were not predicted by the engineers' model of the machine. Similarly, the control systems interact with the human systems surrounding them, creating further sources of error. Poorly designed console systems, faulty maintenance, false alarms, and signals that convey contradictory information can produce worker responses that amplify machine failures.

Finally, failures occur when managers modify the machines to respond to changing product demands and product specifications. The new machine systems create the very conditions that strain their operation. Because machine systems are flexible, the rate of product innovation increases, while automatic controls enable machines to produce to tighter specifications. As the demand for both quality and new products increases, workers and engineers must continually modify the controls. The machines do not fail completely; rather, they require continual upgrading and redesign to remain valuable.

The problem of nuclear power plant design and operation typifies these new challenges. The failures at Three Mile Island, Indian Point, and the Ginna nuclear reactor are signs that we have not yet learned to design machine systems that are integrated with effective job systems. We can no longer think of the worker as simply redundant. Rather, we must design jobs in such a way that workers can effectively control the controls, modifying them and regulating them to prevent failures and errors unanticipated by the engineers. To do so, we must transcend our Taylorist inheritance and develop a new theory and practice of job design.

The engineering response to these emerging problems has been limited. After the Three Mile Island accident, managers and engineers in the nuclear power industry demonstrated great interest in the "human factors" approach to machine design. Human factors engineers examine the role of the individual worker as he focuses his attention, retrieves information, and makes decisions. Such a narrow approach perpetuates the Taylorist emphasis on the single job or task. In post-industrial settings, where workers face uncertain, novel, or dangerous situations, group behavior can shape an individual worker's responses to the machinery. The very character of postindustrial work, of second-order control tasks, of monitoring and evaluating signals and data, increases the significance of group processes. Yet managers and engineers continue to make work-design decisions as if group life did not exist.

To give a recent example, after the mishap at Three Mile Island the Nuclear Regulatory Commission ordered utilities to place senior operators in "shift technical adviser" positions at each operational site. These advisers would be either on call or on site to help workers with difficult technical problems. Informal reports suggest that these advisers have been well received by workers. Yet given the nature of group processes, such an arrangement may prove dangerous. Workers who control complicated machinery feel anxious. To relieve their anxiety, the workers may cede to the adviser their responsibility for vigilantly monitoring the system. The adviser's presence can make them lax. As workers become dependent on him, the number of problems or unforeseen accidents may actually increase, paradoxically affirming the need for the adviser. The work group's relationship to the machinery will be mediated by its relationship to the adviser.

The Problem of Quality Assurance
The process of abdication of responsibility limits the effectiveness of safety inspectors in nuclear power plants.[17] The safety inspectors, called quality engineers, review operators' work procedures, and the certifying documents that record these procedures, in order to ensure that work is done safely. At most nuclear power plants quality engineers work within a distinct quality-assurance (QA) department. They do not report to line managers in the plant but to a QA supervisor. They are thus not part of the operating team.

This has three consequences. First, since they are not part of the operating team, quality engineers often lack concrete knowledge of plant problems; often they do not know where critical safety problems are emerging and which sites are most vulnerable. Second, operating personnel often feel that quality engineers lack technical understanding; therefore they resent the quality engineers' meddling. Third, and most important, operating personnel abdicate their responsibility for working in a safe manner and for assuring other units and divisions that safe procedures have been followed. Safety practices become identified with the meddlesome and ill-informed quality engineer. These practices become *his* business.

Indeed, since the potential dangers are so great at nuclear power plants, operators are psychologically prone to discount their responsibility. The quality engineer is a convenient repository for feelings of danger: the operator unconsciously says, "Let the quality engineer worry about it." This transfer or projection of responsibility serves to reduce the operator's anxiety in two ways: first, if the quality engineer worries, then the operator need not; second, if safety has been entrusted to the quality engineer, with his limited understanding, then things cannot really be so dangerous after all!

These processes of abdication and projection decisively shape the behavior of the quality engineers. When they audit the practices of a particular unit and encounter hostility, they try to avert conflicts by systematically avoiding the investigation of controversial practices. For example, a valve may not be officially termed critical to safety (it is in a building adjoining the containment and does not directly control the flow of cooling water), but a past accident suggests that it may nonetheless be "related to safety." Maintenance work on the valve may be irregular and poorly documented, but the maintenance manager can and will forbid the auditor to inspect records pertaining to the purchase and maintenance of the valve. To minimize conflict, quality engineers focus on areas where minimal judgment is required ("document X was not signed by the shift supervisor as it should have been") so that they can defend their audit findings to both the plant manager and the quality-assurance supervisor.

Indeed, the quality-assurance supervisor may tacitly support such a practice. Because he needs to defend the quality-assurance function against the hostility of line management, he wants to minimize his own conflicts with higher-level management. He will instruct his subordinates to "solve conflicts where they occur"—on the line with operating personnel. But since the quality engineer feels unsupported and unwelcome on the line, this injunction translates into "don't make waves."

The supervisor's message leads the quality engineer to further limit the scope and relevance of his audit. He starts to act very bureaucratically and defensively, attending only to practices that are specifically regulated by Nuclear Regulatory Commission codes. He does not examine each practice as part of a pattern, as a marker of a climate of safety, or as an indicator of likely trends and future problems. This posture only confirms the operators' belief that the quality engineer is a bureaucrat with little understanding of real technical issues.

Finally, this system of interlocking behaviors keeps information about safety problems away from line managers. Because the quality engineers are not making waves, senior managers interpret the absence of conflict between quality engineers and operating personnel as a sign that everything is okay.

The intergroup dynamic described here is rooted in an organizational design that severs the responsibility for safety from the responsibility for operations. The Nuclear Regulatory Commission mandates that safety audits be conducted by inspectors who are independent of the line. But as one nuclear engineer told me, this does not mean that daily safety reviews and the continual refinement of safety practices should be the responsibility of a separate unit. To give an analogy, a company financial controller periodically reviews each division's budget, but each division is in turn responsible for monitoring and controlling its own payables and receivables. The division accountant reports directly to the division manager, not to a jurisdictionally separate budget department. The prevailing split between safety and operations in nuclear power plants can partly be explained by utility management's ignorance of quality control and its resentment of government regulation. But the split is also caused by the psychological processes of abdication and projection that I have already described. The very character of postindustrial work — the uncertainty and possible danger associated with second-order control tasks — makes work groups vulnerable to these psychological processes. Thus intergroup dynamics are both a consequence and a cause of the basic organizational design. The design reinforces the dynamics, and the dynamics reinforce the design.

Remarkably, neither utility engineers nor Nuclear Regulatory Commission engineers understand the interlocking behaviors and assumptions that hobble quality-assurance departments. They do not see how the quality engineer's relationship to his skills, tools, and knowledge is mediated by group processes. Instead, specialists teach quality engineers how to defuse hostility, how to act professionally under fire, how to write more relevant and synthetic reports. Such training reflects engineers' belief that technical fixes can solve human problems. With such fixes engineers try to short-circuit the system of relationships that creates the problem. But technical fixes rarely succeed, because the system of relationships is resistant to change. If the quality engineer acts coolly to dispel hostility, the line supervisor will interpret this behavior as a sign that the engineer finally understands his role. If the quality engineer writes more synthetic reports, the line supervisor will argue that the quality engineer is overstepping his bounds, that he should report discrete findings and stick to the facts.

The technical fix as a solution reflects a human factors approach to understanding human behavior in technical systems.

The Multidisciplinary Design Team

Engineers as a group will not become social psychologists. But there is evidence that they can learn to work in multidisciplinary design teams in which engineering designs are evaluated by professionals, workers, and users who bring different perspectives to the design process. The Big Chem factory, for example, was designed from the start by a multidisciplinary team consisting of company engineers, the plant manager and his staff, representatives from the international union, and a design consultant with a behavioral and technical background.

The team helped engineers avoid two critical errors. Both the union representatives and the management group wanted to design a plant based on a multiskilling system in which workers would perform many different jobs. But the engineers, following traditional practice, proposed to place the laboratory half a mile from the control room. Both management and labor argued that it would prove difficult for workers to test chemicals if the laboratory were so far away and that a specialized laboratory group might emerge in contradiction to the objectives of the design. The laboratory was placed closer to the control room.

Similarly, when engineers were designing the control panels for the machinery, they assumed that the panels would show the customary data on the physical status of variables. Management and labor argued that if workers were to control the machinery, they needed to understand the consequences of their control decisions. Since workers were ultimately responsible for the cost of production, the design group decided to install panel indicators that measured the dollar consequences of various decisions. A shift team that got feedback on the cost impacts of their decisions could more effectively discriminate between good and bad decisions.

The experience at Big Chem suggests that engineers need not become social psychologists to change their narrow, human factors approach. Rather, they must learn the skills that would help them to work in multidisciplinary teams. Engineers need not give up their specialization. Rather, they need to overcome their current isolation and learn to learn from the criticisms and responses of other professionals, users, and workers.

The Discontinuity in the Relationships among People at Work

Postindustrial settings integrate work and learning. The new technology does not deskill workers: they must learn to manage technical transitions from one machine state to another, from expected to unexpected machine events, and from one product design to another. Yet our historical models of skill and the skilled worker are insufficient. The craftsman's skill was too limited in scope, the operator's tacit understanding of the production process lacks theoretical depth, and the engineer often lacks the operator's intuitive grasp of regularities that are hard to measure and assess. Postindustrial skill rests on the integration of these modes of knowing. Workers will develop this integrated understanding only through a process of active learning, direct intervention in the machine system, and progressive widening of their knowledge. The workers' knowledge will have a double character: in solving unstructured and unanticipated problems they will learn how to learn.

The Sociotechnical Setting

As we have seen, settings that integrate work and learning are based on new organizational designs, new management systems, and new modes of interpersonal behavior. The sociotechnical factories established during the last decade are the best working models we have of such settings. In these factories workers are organized into teams; they are paid according to how much they have learned; they participate in peer evaluation; and they become multiskilled. The teams become self-managing. Supervisors, believing that they must teach as well as manage, try not to intervene.

Such systems have particular failure modes. In response to developmental tension, managers may create rules, regulations, and procedures that limit both team autonomy and opportunities for worker learning. Alternatively, managers may abdicate their responsibility, failing to support teams that have not developed the technical capacity to regulate the machines or the social capacity to regulate their own group processes. Workers, for their part, may respond to uncertainty and novelty by creating artificial divisions between teams or workers, thus forming islands of stability and security in the plant community. Alternatively, they may ignore real differences between teams or team members in order to avoid important but difficult conflicts.

While the internal structure of sociotechnical plants gives rise to certain types of failure, workers and managers also fail because these plants are embedded in more inclusive political and social systems.

External forces, such as corporate financial policy, corporate career ladders, the role of unions, and broader social policies, may impinge on the new work community and ultimately hobble it. With such forces operating, workers and managers must frequently rely on their interpersonal skills alone to resolve conflicts. These skills are stretched thin when workers and managers fail to win the support of other actors, institutions, and policies.

Five Modes of Failure

The Role of Unions. The new work setting can fail when workers' inability to manage their own conflicts forces management to take over and reduce team autonomy. Here unions can play a role. Most of these settings are not unionized, but unions can provide a counterstructure to the plant design, within which workers can address their own internal conflicts without calling upon management. Thus, for example, Big Chem continues to develop and evolve as a new work setting in part because the union members have learned to handle worker-to-worker grievances through union committees.[18] Unions have historically not only provided a vehicle for addressing worker-management differences. but also enabled workers to resolve their own differences and to regulate their own internal relationships. Yet management hostility to unions, as well as union leaders' own reluctance to engage in work-design projects, forces workers to rely on their interpersonal skills alone. If these skills prove insufficient, they must then turn to management. They lack a supporting counterstructural frame.

The Economics of Time. If the new work settings are to function at their optimum level, work groups and managers must make mistakes. Senior managers will support a learning climate, where failures can be expected, if they do not feel pressured to show immediate success. Here the economics of time plays a critical role. If senior managers of the parent company reward short-term profit performance, then the plant managers cannot forgive work-group errors. Machine downtimes or high reject rates cannot be ignored in the interest of team and worker learning. The manager may feel compelled to limit team autonomy by reintroducing first-line supervisors as soon as possible.

But the economics of time is shaped by the habits and practices of financial institutions. In the American economy many companies grow by issuing equity rather than debt. American companies' equity value is about three times the value of their debt.[19] Since stock market prices change on the basis of fluctuations in quarterly earnings, top managers must make decisions on the basis of such fluctuations. If, in addition, top management focuses largely on financial indicators of corporate performance (as opposed to indicators of productivity, quality, and market share), stock price changes play an even more central role in regulating production, investment, and divestiture decisions. Finally, when a company is actually a conglomerate of many different units producing various goods and services, financial indicators are often the only mechanism for integrating and assessing the performance of the conglomerate as a whole. Thus a plant manager's local decision to introduce a supervisor is shaped by broad and deeply rooted financial practices.

Management Teams. Managers must understand team dynamics, intragroup collaboration, and learning processes if they are to manage the new work settings effectively. Yet the structure of management careers in large American companies rarely rewards collaboration between managers themselves. Managers are promoted because they have individually excelled. Managers on the fast track are shifted frequently from one unit to another to give them exposure. But as a consequence they rarely develop close ties to a group of their peers. In many large companies ambitious managers resist joining special projects, such as building and staffing a new facility, because these involvements take them away from company headquarters and reduce their exposure. But it is precisely on such multidisciplinary projects— which often cut across divisions and units—that managers learn to work in teams.

The same career dynamics limit managers' ability to understand how groups learn, particularly how they learn from failures. I once worked with the top management group of a plastics company to help them improve their team dynamics. The project failed. The president was very domineering and had strong individual relationships with the various vice-presidents, but the group as a whole had a poorly articulated team process. The vice-presidents mistrusted one another, each suspecting that the others had hidden agendas. At one point I asked them if they systematically reviewed the bad business decisions they had made. They said no, even though in the past year they had made a bad decision to produce a synthetic material for an industrial container that had proved to be environmentally dangerous. A post-mortem would have helped them understand the problems they faced in coordinating the work of their research chemists and production managers.

Their poor team process was integrally connected to their unwillingness to review past failures. When a group reviews its failures, people must expose the judgments they made that contributed to the bad decision. Group members will take such risks only if they trust one another—only if they believe that in assessing their personal contributions to the failure they will not expose themselves to retribution. A group whose members refuse to engage in this process cannot collectively learn from important past failures—each member will have his own personal theory that can never be tested—and cannot learn how to learn. Under these conditions management cannot understand or effectively support the new work settings.

Role Specialization. We have seen how the new plants fail when workers refuse to rotate among divisions or units, often splitting the work force into separate skill groupings. A split between warehouse and line workers developed at Fall Mills when the former refused to rotate onto the line. Anxiety plays a part: workers fear they cannot master all the skills. But this anxiety is compounded by the tradition of specialization that pervades our economy. By defining it as specialized or difficult to learn, a group of workers can control entry to a job, craft, or profession. This is how craft trades and professions defend their monopolies. Career security and mobility depend on specialization. Such specialization in turn impedes workers' abilities to diagnose and solve machine failures in which electrical, hydraulic, and pneumatic systems are all interrelated.

Social Policy. In a sociotechnical system all workers can reach the top rate; those who wish to continue to grow must look beyond the factory. But when a worker leaves the factory in search of a more demanding job, he risks being unemployed. The risk may induce him to stay, even though he feels he is stagnating. If many senior workers decide to stay, a rift may develop between the old-timers and the newcomers. The former will be less interested in teaching because they have fewer opportunities to learn. Managers anxious to achieve some optimum level of turnover in order to sustain the factory's learning climate—and limit the wage bill—may then pressure some workers to leave.

This difficulty can be eased by social policies that reduce the risks and costs of job transitions. Analysts have described the scope and structure of such policies: pensions should be portable, people should be allowed a fixed amount of free schooling over their entire lives, and retraining programs should be expanded. Such programs would facilitate job transitions by reducing the risks for workers and the cost pressures on companies. But if opposition to government spending continues to dominate American domestic policy at both the state and federal levels, it may not be possible to develop and fund such programs; the incipient tension between senior and junior workers, between designs that facilitate learning and designs that reinforce the traditions of craft and expertise, may be exacerbated.

Facilitating Forces

Career dynamics, financial pressures, social policies, the role of unions, and the culture of specialization all shape the emerging relationships between people in the new work settings. Again, it is easy to be pessimistic here. I expect that the dynamics of management careers will prove central: unless managers themselves learn to work in teams and are promoted for the quality of their teamwork, it is hard to see how settings based on shop-floor teams can function successfully.

Nonetheless, several interdependent processes are now taking place that may restructure middle-management careers and increase managers' commitment to teamwork. First, in contrast to the situation in previous recessions, the number of managers and the size of their staffs relative to the size of line divisions is falling in many companies.[20] Companies facing competition from new products and technologies, often within newly deregulated markets, are trying to make permanent reductions in management and overhead costs. Since managers have smaller staffs, they find it harder to build up personal departmental fiefdoms.

Next, since the management span of control is rising, managers must delegate a greater number of their tasks downward, thus increasing the autonomy of those below them. In addition, because managers have smaller staffs and budgets, they can no longer easily buffer their departments from the decisions of other departments. For example, if the Management Information Systems division of company X refuses to sign off on division Y's request to buy a certain computer system, division Y has fewer dollars and employees to short-circuit the official decision process and buy its own. This means that the manager of division Y must develop better ties with other division managers if he is to get the resources he needs.

As the number of management positions falls, managers may be able to excel only by accepting lateral transfers and demonstrating competence in many divisions. These lateral transfers would no longer be short stops on the way up, but instead part of a permanent system of movement between divisions, perhaps including several returns to each division. Lateral transfers may create a climate in which managers value special-project work. The skills required to transfer laterally with success are similar to those required to manage interdivisional projects. In each case the manager must understand the cultures and perspectives of various divisions. Thus, as vertical promotions become less frequent, interdepartmental teamwork may grow.

Finally, as competition increases, companies must return to production basics. They may no longer be able to rely on the financial strategies of divestiture, merger, and acquisition to increase their profitability. These strategies increasingly fail as both deregulation and new technologies make the future value of a given company hard to predict. Instead, senior managers must focus on shop-floor problems, production management, and quality control. In so doing, they will reward managers who develop long-term improvement programs and who stay with a particular division until a program is completed.

All these processes, if they take hold, will reshape middle-management careers in ways that will increase managers' incentives to support teamwork and work-group autonomy at all levels. Predictions are suspect here; yet current company cutbacks in management staff reflect in part the postindustrial transition itself. Companies are cutting back partly in response to the uncertainty and failures induced by the new technologies. The microprocessor is changing the productivity differential between companies, while the computer is restructuring the communications complex, the banking industry, education, printing, and publishing. If these processes of competition are sustained, the postindustrial transition will squeeze companies from two sides. In response to market uncertainties senior management will promote management teams, and in response to the new technologies senior management will promote worker teams. This is a possible, if not certain, path of development to a postindustrial society.

In Sum

If we are to design work settings that will help us make full use of the new technologies, we must confront the constraints imposed by broader social forces. The piece-rate system of machine shops and management's withdrawal from the shop floor create significant obstacles to appropriating the new machine-tool technology. Engineers' resistance to the study of behavior, their human factors approach to technical design, and their unwillingness to consider the role of group processes at work all ultimately hobble the worker-machine systems they design. Finally, the career dynamics of middle management, the ethos of professionalization, and the use of short-term planning horizons in determining company profits all limit management's capacity and willingness to support teamwork on the shop floor.

Certain enabling forces are emerging. Workers' capacity to manage themselves (as revealed in their games on the shop floor), engineers' proven ability to work in multidisciplinary teams, and the reduction in the size of management will all help engineers, workers, and managers appropriate the new technologies. But we may not be able to rely on these forces alone. Their very diversity suggests that we may have to reconstruct the institutional framework within which companies, governments, unions, and schools develop, implement, and coordinate their plans.

The new technologies do not constrain social life and reduce everything to a formula. On the contrary, they demand that we develop a culture of learning, an appreciation of emergent phenomena, an understanding of tacit knowledge, a feeling for interpersonal processes, and an appreciation of our organizational design choices. It is paradoxical but true that even as we are developing the most advanced, mathematical, and abstract technologies, we must depend increasingly on informal modes of learning, design, and communication.

The new factory settings will fail in many ways as they appropriate the new technologies. But we can learn from postindustrial practice itself. In postindustrial settings we learn by analyzing failures. Indeed, failures are the precondition for both work and learning. We must view current failures, not as signs of impossibilities, but as indicators of the unique problems we face in managing the transition to a postindustrial work system. Only in this way can we develop designs and policies that will take us beyond mechanization.

Notes

Introduction

1. Sigfried Giedion, *Mechanization Takes Command: A Contribution to Anonymous History* (New York: W. W. Norton, 1948), 121.

2. Louis E. Davis and Charles S. Sullivan, "A Labour-Management Contract and the Quality of Working Life," *Journal of Occupational Behavior* 1 (1980), 34.

3. From *Good Work Practices*, handbook of the chemical plant.

4. Davis and Sullivan, "A Labour-Management Contract," 32.

Chapter 1

1. Sigfried Giedion, *Mechanization Takes Command*.

2. For a general history of the lathe and other machine tools see K. R. Gilbert, "Machine Tools," in *A History of Technology*, ed. Charles Singer (Cambridge: Cambridge University Press, 1958), vol. 4.

3. Robert S. Woodbury, "Machines and Tools," in *Technology in Western Civilization*, ed. Melvin Kranzberg and Caroll Pursell, Jr. (London: Oxford University Press, 1967), 1:621.

4. Ibid., 623.

5. Ibid., 624–625.

6. Robert S. Woodbury, "The Legend of Eli Whitney and Interchangeable Parts." in *Technology and Culture*, ed. Melvin Kranzberg and William Davenport (New York: Schocken Books, 1972).

7. Giedion, *Mechanization Takes Command*, 77n.

8. Ibid., 77–78.

9. James R. Bright, *Automation and Management* (Cambridge: Harvard University Press, 1958), 243, Appendix 3.

10. Ibid.

11. Daniel Nelson, *Managers and Workers: The Origins of the New Factory System in the United States, 1880–1920* (Madison: University of Wisconsin Press, 1975), 17–33.

12. Ibid., chap. 4.

13. Bright, *Automation and Management*, 22–30.

14. Abbot Payton Usher, *A History of Mechanical Inventions* (Cambridge: Harvard University Press, 1954), 116.

15. Bright, *Automation and Management*, 67–69.

16. For the classic criticism of Taylorism see Georges Friedmann, *Industrial Society* (New York: Free Press, 1964).

17. Frederick Winslow Taylor, *The Principles of Scientific Management* (New York: W. W. Norton, 1967).

18. Giedion, *Mechanization Takes Command*, 104.

19. "Symposium: Stopwatch Time Study: An Indictment and Defense," *Bulletin of the Taylor Society*, June 1921, 109.

20. Giedion, *Mechanization Takes Command*, 115.

21. Ibid., 721–723.

Chapter 2

1. Marx makes a somewhat different trichotomy. He writes, "All fully developed machinery consists of three essentially different parts, the motor mechanism, the transmitting mechanism, and finally the tool or working machine." *Capital* (New York: International Publishers, 1967), 1:373.

2. Nathan Rosenberg, *Perspectives on Technology* (Cambridge: Cambridge University Press, 1976), 307 n. 28.

3. Ibid., 118.

4. D. F. Galloway, "Machine Tools," in *A History of Technology*, 5:640–641.

5. Malcolm Malloren, *The Rise of the Electrical Industry during the Nineteenth Century* (Princeton: Princeton University Press, 1943), 90–92.

6. Ibid.

7. William E. Leuchtenberg, *The Perils of Prosperity: 1914–1932* (Chicago: University of Chicago Press, 1958), 179.

8. John H. Lorant, *The Role of Capital-Improving Innovations in American Manufacturing during the 1920s* (New York: Arno Press, 1975), 118.

9. Ibid., 129–130.

10. George H. Amber and Paul S. Amber, *Anatomy of Automation* (Englewood Cliffs, N. J.: Prentice-Hall, 1962), 146.

11. Marx, *Capital*, 385.

12. Woodbury, "Machines and Tools," 626.

13. *The Way Things Work* (New York: Simon and Schuster, 1971), 212–213.

14. Ibid., 629.

15. Ibid., 283, fig. 6.

16. Taylor, *Scientific Management*, 97–113.

Chapter 3

1. Norbert Wiener, *God and Golem* (Cambridge: MIT Press, 1966), 32.

2. Stafford Beer, *Cybernetics and Management*, (New York: John Wiley, 1959), 29.

3. Otto Mayr, *The Origins of Feedback Control* (Cambridge: MIT Press, 1970), 107.

4. Ibid., 112.

5. Ibid., 112–113.

6. Ibid., 131.

7. Ibid.

8. James C. Maxwell, "On Governors," *Proceedings of the Royal Society* 16 (1867–68): 270–283.

Chapter 4

1. R. Kompfher, "Electron Devices in Science and Technology," in *Turning Points in American Electrical History*, ed. James E. Brittain (New York: The Institute of Electrical and Electronics Engineers Inc., 1977), 333.

2. *Process Integration and Instrumentation*, (London: British Electrical Development Association, 1959), 41.

3. J. A. Fleming, *Fifty Years of Electricity* (London: The Wireless Press, 1921), 345.

4. Norbert Wiener, *I Am a Mathematician* (Cambridge: MIT Press, 1964), 72.

5. *Modern Men of Science*, ed. Jay Greene Elihu, s.v. "Black, Harold Stephen" (New York: McGraw-Hill, 1966), 42.

6. Statement by M. J. Kelly, cited in the introduction to H. S. Black's paper, "Stabilized Feedback Amplifiers," in *Turning Points in American Electrical History*, 342.

7. Nic J. T. A. Kramer and Jacob de Smit, *Systems Thinking* (Leiden: Martinus Nijhoff, 1977), 3.

8. Ibid.

9. Norbert Wiener, *Cybernetics* (Cambridge: MIT Press, 1948), chap. 1.

10. D. S. L. Cardwell, *Turning Points in Western Technology* (New York: Neale Watson Academic Publications, 1972), 51–53.

11. Ibid., 92.

Chapter 5

1. Giedion, *Mechanization Takes Command*, 41.

2. John H. Lorant, *The Role of Capital-Improving Innovations in American Manufacturing during the 1920s* (New York: Arno Press, 1975), 63.

3. *Encyclopaedia Britannica*, 15th ed., s. v. "chemical engineering."

4. Lorant, *Capital-Improving Innovations*, 86.

5. Lorant, *Capital-Improving Innovations*, 100.

6. Eugene Ayres, "An Automatic Chemical Plant," *Scientific American*, September 1952, 91.

7. Ibid., 94.

8. Lorant, *Capital-Improving Innovations*, 187.

9. For example, in this journal we find William A. Peters, Jr., "Automatic Fractionating Columns," June 22, 1923; W. Trinks, "The Control of Furnace Atmosphere," December 4, 1924; Harold C. Webber, "The Full Electrode Vacuum Tube and Its Application to Some Chemical Engineering Problems," June 1, 1927; John J. Grebe, "The Control of Chemical Processes" (special emphasis on sensing devices), 1930; W. R. King, "Electron Tubes—Their Industrial Application," December 1932.

10. *McGraw-Hill Encyclopedia of Science and Technology*, s. v. "petrochemical" (New York: McGraw-Hill, 1971), 10:51.

11. *Encyclopaedia Britannica*, 15th ed., s.v. "chemical industry."

12. Lorant, *Capital-Improving Innovations, 106.*

13. Larry Hirschhorn, "The Theory of Social Services in Disaccumulationist Capitalism," *International Journal of Health Services* 9, no. 2 (1979), 299.

14. Aubrey Burstall, *A History of Mechanical Engineering* (Cambridge: MIT Press, 1965), 381.

15. See, for instance, the general text by Lawrence E. Doyle, *Metal Machining* (Englewood Cliffs, N.J.: Prentice-Hall, 1953).

16. The following material is drawn from John E. Ward, "Numerical Control of Machine Tools," *McGraw-Hill Yearbook of Science and Technology, 1968,* (New York: McGraw-Hill, 1968); and William Pease, "An Automatic Machine Tool," *Scientific American,* September 1952.

17. Melvin Blumberg and Donald Gerwin, "Human Consequences Relating to the Acquisition and Use of Computerized Manufacturing Technology" (paper presented at Quality of Working Life Conference, Toronto, August 1981), 2.

18. Donald Gerwin, "Dos and Don'ts of Computerized Manufacturing," *Harvard Business Review,* March 1982, 107–116, esp. 108.

19. Giedion, *Mechanization Takes Command,* 118–120.

20. Gerwin, "Dos and Don'ts," 112.

21. Ibid., 113–114.

Chapter 6

1. The argument that follows is drawn from Raymond J. Krekel and John B. Loague, "Cement Plant Computer—Success or Failure," *IEEE Transactions on Industry and General Applications,* September–October 1971, 602–609.

2. Clyde W. Moore, "A Survey of X-Ray Analyzers as Applied in the Cement Industry," *IEEE Transactions in Industry and General Applications,* November–December 1977, 563–564.

3. Krekel and Loague, "Cement Plant Computer," 606.

4. This section is based on personal communication with Dr. Daniel Perlmutter, Department of Chemical Engineering, University of Pennsylvania, July 1982.

5. James L. Nevins, "Sensors for Industrial Automation," *McGraw-Hill Yearbook of Science and Technology, 1975* (New York: McGraw-Hill, 1975), 60.

6. Ibid.

7. David Noble, "Social Choice in Machine Design: The Case of Automatically Controlled Machine Tools," in *Case Studies on the Labor Process,* ed. Andrew Zimbalist (New York: Monthly Review Press, 1979), 40–41.

8. Robert Wesson, Kenneth Salomon, Robert Steeb, Perry Thorndyke, and Kenneth Westcourt, *Scenario for the Evolution of Air Traffic Control* (Santa Monica: Rand, 1981), 42–44.

9. Blumberg and Gerwin, "Human Consequences," 11.

10. Fred Emery, "The Fifth Wave: Embarking on the Next Forty Years" (mimeo, Canberra, May 1978), 16.

11. Robert Boguslaw, *The New Utopians: A Study of System Design and Social Change* (New York: Prentice-Hall, 1965).

12. Federico Butera, "Environmental Factors in Job and Organization Design: The Case of Olivetti," in *The Quality of Working Life,* ed. Louis E. Davis and Albert B. Chearns (New York: Free Press, 1975), 2:166–200.

Chapter 7

1. Harry Braverman, *Labor and Monopoly Capital* (New York: Monthly Review Press, 1974).

2. Daniel Nelson, *Managers and Workers: Origins of the New Factory System in the United States* (Madison: University of Wisconsin Press, 1975), chap. 3.

3. Ibid., 40.

4. Ibid., 41.

5. Ibid., 49.

6. Ibid., 69.

7. Ibid., 75.

8. Georges Friedmann, *Industrial Society* (New York: Free Press, 1964), pt. 2, chap. 2.

9. Elton Mayo, "What Is Monotony?" in *Modern Technology and Civilization*, ed. Charles Walker (New York: McGraw-Hill, 1962).

10. Ibid., 86.

11. Ibid., 88–89.

12. Friedmann, *Industrial Society*, pt. 3, chap. 3.

13. Georges Friedmann, *The Anatomy of Work* (New York: Free Press, 1961), chaps. 3 and 4.

14. William Whyte, "An Experiment in Worker Control over Pacing," in *Modern Technology and Civilization*.

15. The description of this plant is drawn in its entirety from Charles Walker, *Toward the Automatic Factory* (New Haven: Yale University Press, 1957).

16. R. G. Sell et al., "An Ergonomic Method of Analysis Applied to Hot Strip Mills," *Ergonomics*, 1962, 203–211.

17. Walker, *Toward the Automatic Factory*, 171–172.

18. Ibid., 173.

19. Ibid., 90.

20. Floyd C. Mann and L. Richard Hoffman, *Automation and the Worker* (New York: Henry Holt and Company, 1960).

21. Otis Lipstreau and Kenneth A. Reed, *Transition to Automation*, Series in Business, no. 1 (Boulder: University of Colorado Press, 1964).

22. Robert Blauner, *Alienation and Freedom*, (Chicago: University of Chicago Press, 1964) chaps. 6 and 7.

23. Braverman, *Labor and Monopoly Capital*, 213–223.

24. Bright, *Automation and Management*.

25. Braverman, *Labor and Monopoly Capital*, 114.

26. Walker, *Toward the Automatic Factory*, 31.

27. Lipstreau and Reed, *Transition to Automation*, 63.

28. Walker, *Toward the Automatic Factory*, 161–167.

Chapter 8

1. Mike Gray, "What Really Happened at Three Mile Island," *Rolling Stone*, May 17, 1979, 47.

2. The sequence of the accident is drawn from Office of Nuclear Reactor Regulation, "Staff Report on the Generic Assessment of Feedwater Transients in Pressurized Water Reactors Designed by the Babcock and Wilcox Company" (Washington, D.C.: U. S. Nuclear Regulatory Commission, 1979); The President's Commission, *The Accident at Three Mile Island* (Washington, D. C.: Government Printing Office, 1979); and Gray, "What Really Happened."

3. "Staff Report on the Generic Assessment," 2–32.

4. Gray, "What Really Happened," 48.

5. "Staff Report on the Generic Assessment," 1–5.

6. Gray, "What Really Happened," 48.

7. "Staff Report on the Generic Assessment," 2–27.

8. Sir Arnold Lindley and Sir Stanley Brown, "The UKsmouth 60 MW Turbine Failure," in *Engineering Progress through Trouble*, ed. R. R. Whyte (London: Institution of Mechanical Engineers, 1975).

9. Ralph Blumenthal, "How Water Leak Shut Down Con Edison's Reactor," *New York Times*, November 3, 1980, B-3.

10. *New York Times*, September 21, 1981.

11. Henry Petroski, "When Cracks Become Breakthroughs," *Technology Review*, August–September 1982, 24–25.

12. O. M. Smith, "Vibrations in Turbo Machinery," in *Engineering Progress through Trouble*, 79–84.

13. J. P. Den Hartog, "Vibrations—A Survey of Industrial Applications," in *Engineering Progress through Trouble*, 86.

14. Petroski, "When Cracks Become Breakthroughs," 20.

Ibid., 21.

16. Lindley and Brown, "The UKsmouth 60 MW Turbine Failure," 68.

17. Joseph R. Egan, "To Err Is Human Factors," *Technology Review*, February–March 1982, 23–29, esp. 24.

18. Karl L. Wiener, "Controlled Flight into Terrain Accidents: System-Induced Errors," *Human Factors* 19, no. 2 (1977), 171–181.

19. Robert Trivers and Huey P. Newton, "The Crash of Flight 90: Doomed by Self-Deception?" *Science Digest*, November 1982, 66.

20. John E. Robinson, Walter E. Deutsch, and James E. Rogers, "The Field Maintenance Interface between Human Engineering and Maintainability Engineering," *Human Factors* 12, no. 2 (1970), 253–259, esp. 256–258.

21. Ibid., 258.

22. Richard Witkin, "FAA Reports Disturbing Faults in Two DC-10s and Asks New Checks," *New York Times*, June 14, 1979, A-14.

23. Joel Greenberg, "Human Error," *Science News* 117 (February 23, 1980), 125.

24. Ibid.

25. Alphonse Chapanis, "Theory and Methods for Analyzing Errors in Man-Machine Systems," in *Selected Papers on Human Factors in the Design and Use of Control Systems*, ed. Wallace A. Sinaiko (London: Dover, 1961), 53.

26. Themis P. Speis, "Preliminary Evaluation of Operator Actions for SG Tube Rupture Event," memorandum, Nuclear Regulatory Commission, January 28, 1982, 2.

27. Sir G. de Havilland and P. B. Walker, "The Comet Failure," in *Engineering Progress through Trouble*, 53.

28. Frederick P. Brooks, Jr., *The Mythical Man-Month* (Reading, Mass.: Addison-Wesley, 1979), 121.

29. Andrew Pollack, "Old Programs, New Problems," *New York Times*, November 24, 1981, D-1.

30. Thomas B. Sheridan, "Human Error in Nuclear Power Plants," *Technology Review*, February 1980, 28.

31. President's Commission, *The Accident*, 96.

32. *New York Times*, July 14, 1980, A-16.

33. William Robbins, "Testimony Suggests Flaw in Helicopter on Iran Raid," *New York Times*, December 9, 1980, A-16.

34. Shoshanah Zuboff, "Problems of Symbolic Toil," *Dissent*, Winter 1982, 57.

35. President's Commission, *The Accident*, 96, 102.

36. Sussana Jaffee, "All the World Is a Laboratory for Nuclear Technology," *In These Times*, November 19–25, 1980, 10.

37. "Report on Ginna Plant Accident Says Operators Acted Too Late," *New York Times*, February 6, 1982.

38. F. R. Mynatt, "Nuclear Reactor Safety Research since Three Mile Island," *Science*, April 9, 1982, 131.

39. Nicholas A. Bond, Jr., "Some Persistent Myths about Military Electronics Maintenance," *Human Factors* 12, no. 2 (1970), 241–252, esp. 243.

40. Ibid., 244.

41. Hugh M. Bowen, "The Imp in the System," in Elwyn Edwards and Frank P. Lees, *The Human Operator in Complex Systems*, (London: Taylor and Francis, 1967), 15.

42. Council for Science and Society, *New Technology: Society Employment and Skill* (London, 1981), 41.

Chapter 9

"Staff Report on the Generic Assessment," 2–32.

2. Ibid., 2–27.

3. Ibid., 229–239.

4. W. T. Singleton, "Theoretical Approaches to Human Error," *Ergonomics* 16, no. 6 (1976), 727–736, esp. 730.

5. Karl L. Wiener, "Controlled Flight," 178.

6. Gray, "What Really Happened," 48.

7. James Floyd's testimony was broadcast on radio on May 31, 1979.

8. President's Commission, *The Accident*, 94.

9. Ibid., 99.

10. "Staff Report on the Generic Assessment," 4-2.

11. P. M. Haas and T. F. Bott, *Criteria for Safety-Related Nuclear Plant Operator Actions: A Preliminary Assessment of Available Data* (Oak Ridge: Oak Ridge National Laboratory, 1979), 32.

12. Speis, "Preliminary Evaluation of Operator Actions."

13. Dan Clawson, *Bureaucracy and the Labor Process* (New York: Monthly Review Press, 1980), 139–141.

14. F. R. Brighton and L. Laois, "Operator Performance in the Control of a Laboratory Process Plant," *Ergonomics* 18, no. 1 (1975), 63–66, esp. 64; A. Shepherd et al., "Diagnosis of Plant Failures from a Control Panel: A Comparison of Three Training Methods," *Ergonomics* 20, no. 4 (1977), 347–361.

15. Wesson et al., *Evolution of Air Traffic Control*, 24.

16. Haas and Bott, *Operator Actions*, 88.

17. Herbert L. Dreyfus, *What Computers Can't Do: The Limits of Artificial Intelligence* (New York: Harper and Row, 1979), 108–119.

18. President's Commission, *The Accident*, 107.

19. J. Rasmussen, "On the Communication between Operators and Instrumentation in Automatic Process Plants," in *The Human Operator in Process Control*, ed. Elwyn Edwards and Frank P. Lees, (London: Taylor and Francis, 1974), 226–227.

20. J. Rasmussen and A. Jensen, "Mental Procedures in Real-Life Tasks: A Case Study of Electronic Trouble Shooting," *Ergonomics* 17, no. 3 (1974), 293–307.

21. Ibid., 299.

22. K. D. Duncan and A. Shepherd, "A Simulator and Training Technique for Diagnosing Plant Failures from Control Panels," *Ergonomics* 18, no. 6 (1975), 627–641, esp. 638.

23. Gregory Bateson, "Social Planning and the Concept of Deutero-Learning," in *Steps to an Ecology of Mind* (New York: Ballantine, 1972).

24. Haas and Bott, *Operator Actions*, 89.

25. Davis and Sullivan, "A Labour-Management Contract," 34.

26. Charles R. Kelley, *Manual and Automatic Control* (New York: John Wiley, 1967), chaps. 14 and 15.

27. J. C. Jones, "The Designing of Man-Machine Systems," in *The Human Operator in Complex Systems*, 8.

28. Wesson et al., *Evolution of Air Traffic Control*, 36.

Chapter 10

1. Frank P. Lees, "Research on the Process Operation," in *The Human Operator in Process Control*, 393.

2. Per H. Engelstad, "Socio-Technical Approach to Problems of Process Control," in *Design of Jobs*, ed. Louis E. Davis and James C. Taylor (Harmondsworth: Penguin Books, 1972), 338.

3. Derek W. E. Burdew, "Participative Management as a Basis for Improved Quality of Jobs: The Case of Microwax Department, Shell U.K., Ltd.," in *The Quality of Working Life*, vol. 2.

4. Federico Butera, "Environmental Factors."

5. Ibid., 167–168.

6. Ibid., 168.

7. Ibid., 186.

8. Ray Wild, *Work Organization: A Study of Manual Work and Mass Production* (London: John Wiley, 1975), chap. 9.

9. Author unknown, "Fiat: Initiatives in the Field of Work Organization" (paper presented at Quality of Working Life Conference, Toronto, August 1981), 6.

10. Ibid., 7.

11. Ibid., 10.

12. Roy Walters, "The Citibank Project: Improving Productivity through Work Design," in *The Innovative Organization: Productivity Programs in Action*, ed. Robert Zager and Michael P. Rosow (New York: Pergamon Press, 1982).

13. H. Bullinger, H. Warnecke, E. Haller, "Effects of Social, Technological, and Organizational Changes on the Labor Design as Shown by the Example of Micro-electronics" (paper presented at Quality of Working Life Conference, Toronto, August 1981).

14. Charles Sabel, *The Division of Labor: Its Progress through Politics* (New York: Cambridge University Press, 1982), 199.

15. Ibid., 211.

16. Ibid., 206–207.

17. Ibid., 208.

Chapter 11

1. Frank Guere, "Why Nuclear Reactors Are Failing: Perhaps It's a Failure in Training," *The Philadelphia Inquirer*, May 20, 1979.

2. Ibid.

3. Ibid.

4. Larry Blair and John Dogetter, "Education, Training, and Work Experience among Nuclear Power Plant Workers" (Oak Ridge Associated Universities, April 1980), 5–7.

5. Ibid., 7.

6. Guere, "Why Nuclear Reactors Are Failing."

7. Cal Pava, "Autonomous Teams: The State of the Art" (mimeo, University of Pennsylvania, 1979), 5–6.

8. From *Good Work Practices*, handbook of Big Chem.

9. Harvey Kolodny and Barbara Dressner, "Linking Arrangements and New Work Designs" (mimeo, Quality of Working Life Centre, Toronto, 1981), 8.

10. Interview with a facilitator, 1982.

11. Interview with a manager, 1982.

12. David Hawk and William Henderson, "A Study of a Self-Management Work System" (mimeo, University of Pennsylvania, 1975), 7–8.

13. Richard Walton, "The Topeka Work System: Optimistic Visions, Pessimistic Hypotheses, and Reality," in *The Innovative Organization*, 269.

14. Kolodny and Dressner, "Linking Arrangements."

15. Eric Trist, "The Evaluation of Socio-Technical Systems as a Conceptual Framework and as an Action Research Program," in *Organizational Design and Performance*, ed. Andrew Van de Ven and William Joyce, (New York: Wiley Interscience, 1981).

16. Fred Emery and Eric Trist, "Human Organizations as Systems," in *Systems Thinking*, ed. Fred Emery, (Harmondsworth: Penguin Books, 1971).

17. Kurt Lewin, *Field Theory in Social Science* (New York: Harper and Row, 1951).

18. Fred Emery, "The Assembly Line: Its Logic and Our Future," in *Sociotechnical Systems: A Sourcebook*, ed. William A. Passmore and John J. Sherwood (San Diego: University Associates, 1978), 344.

19. Louis E. Davis and Eric Trist, "Improving the Quality of Working Life: Socio-technical Case Studies," in *Design of Jobs*, ed. Louis E. Davis and James C. Taylor (Santa Monica: Goodyear, 1979).

20. Ibid.

21. J. Maxwell Elden, "Democracy at Work for a More Participatory Politics: Worker Self-Management Increases Political Efficacy and Participation" (Ph.D. diss., University of California, Los Angeles, 1979), 100.

22. Ibid., 98.

23. Ibid., 225.

24. Davis and Sullivan, "A Labour-Management Contract," 34.

Chapter 12

1. Davis and Sullivan, "A Labour-Management Contract," 32.

2. Richard Walton, "Establishing and Maintaining High-Commitment Work Systems," in *The Organizational Life Cycle*, ed. John Kimberly and Robert H. Miles and Associates (San Francisco: Jossey-Bass, 1980), 235. Walton calls South Mack "Salem" in his article. I have drawn on his article, as well as interviews, for my material on South Mack.

3. Ibid., 234.

4. Ibid.

5. Ibid., 239.

6. Walton, "The Topeka Worker System," 269.

7. Elden, "Democracy at Work," 98.

8. Interview with a consultant from the parent company of Fall Mills, 1982.

9. Walton, "High-Commitment Work Systems," 235.

10. Interview with a plant supervisor, 1982.

11. Ibid.

12. Ibid.

13. Walton, "High-Commitment Work Systems," 236.

14. Interview with a supervisor, 1982.

15. John Witte, *Democracy, Authority, and Alienation in Work* (Chicago: University of Chicago Press, 1980), 119.

16. Ibid., 120.

17. Ibid.

18. Ibid., 122.

19. Interview with a facilitator, 1982.

20. Interview with a manager, 1982.

21. Interview with a facilitator, 1982.

22. Interview with a consultant, 1982.

23. Interview with a manager, 1982.

24. Walton, "High-Commitment Work Systems," 227.

25. Interview with a supervisor, 1982.

26. Interview with a supervisor, 1982.

27. Don Ronchi and William Morgan, "Persisting and Prevailing in Springfield, Ohio" (mimeo, Ohio State University, 1981).

28. Ibid., 13.

29. Interview with a supervisor, 1982.

30. Interview with a supervisor, 1982.

31. Interview with a manager, 1982.

32. Interview with a manager, 1982.

33. Interview with a researcher, 1982.

34. Interview with a manager, 1982.

35. Elden, "Democracy at Work," 239.

36. Ibid., 240.

37. Witte, *Democracy, Authority, and Alienation*, 119.

38. Interview with a facilitator, 1982.

39. Ibid., 199.

40. Interview with a facilitator, 1982.

41. Interview with a facilitator, 1982.

42. Interview with a facilitator, 1982.

43. Witte, *Democracy, Authority, and Alienation*, 123.

44. Interview with a manager and a supervisor, 1982.

45. Ibid.

46. Witte, *Democracy, Authority, and Alienation*, 31.

47. Eric Trist and Charles Dwyer, "The Limits of Laissez-Faire as a Sociotechnical Strategy," in *The Innovative Organization*, 171.

48. Interview with a facilitator, 1982.

49. Interview with a manager, 1982.

50. Walton, "The Topeka Worker System," 269.

51. Pava, "Autonomous Teams," 7.

52. Beth Atkinson, "Power Management in a Team Organization" (Ph.D. diss., University of Alberta, 1980).

53. Interview with Atkinson.

54. Atkinson, "Power Management," 145–49.

55. Ibid., 146–47.

56. Interview with Atkinson.

57. Atkinson, "Power Management," 14.

58. Chris Argyris, *Reasoning, Learning, and Action* (San Francisco: Jossey-Bass, 1982).

59. Atkinson, "Power Management," 150.

60. Interview with a supervisor, 1982.

61. Interview with an internal consultant of the parent company, 1982.

Chapter 13

1. "Tool Machine Companies Are Hopeful despite Slumps," *New York Times*, September 6, 1982.

2. Nathan H. Cook, "Computer-Managed Parts Manufacturing," *Scientific American*, February 1975, 25.

3. Lydia Chavez, "Hanging Tough on Machine Tools," *New York Times*, August 15, 1982.

4. Michael Burawoy, *Manufacturing Consent* (Chicago: University of Chicago Press, 1982), 47.

5. "Die Casters are Struggling to Adapt," *New York Times*, June 28, 1982.

6. Ibid.

7. Ibid.

8. Burawoy, *Manufacturing Consent*, chap. 4.

9. Ibid., chap. 5.

10. U.S. Department of Commerce, "Metal-Working Machinery and Equipment," *Industrial Outlook* (Washington, D.C., 1982), 176–177.

11. Cook, "Computer-Managed," 25–26.

12. Burawoy, *Manufacturing Consent*, 48.

13. Ibid., 67–70.

14. Ibid., 67.

15. Ibid., 68.

16. Ibid.

17. This analysis is based on consulting work I did with a nuclear engineering company.

18. Interview with a researcher, 1982.

19. Personal communication, Professor Jay Ritter, Rodney L. White Center for Financial Research, Wharton School, University of Pennsylvania, 1983.

20. Tom Gilmore and Larry Hirschhorn, "Managing Decline," *Journal of Human Resources*, December 1983.

Index

General Public Utilities, 113. *See also*
Nuclear reactors: Three Mile Island
Giedion, Sigfried, 1, 5, 8, 9, 41, 49
Gilbreth, Frank, 14. *See also*
Management, scientific; Taylor,
Frederick Winslow
Goldberg, Rube, 16, 17
Governor, 31–33. *See also* Steam
engine; Watt, James

Habit, preconception and, 88–90. *See
also* Learning
Hierarchy, coordination without,
150–151. *See also* Design,
sociotechnical
Homeostasis, 39, 156
Household, Inc., 122, 141. *See also*
Design, sociotechnical; Factories,
sociotechnical
Hoxie, Robert, 63

Industrialization. *See* Mechanization
Industrial management. *See*
Management, scientific
Integration, flexibility and, 57–58
Interchangeability. *See* Standardization

Job design. *See* Design

Keller duplication machine, 47
Köhler, Wolfgang, 39

Labor. *See* Workers
Labor and Monopoly Capital, 62. *See also*
Braverman, Harry
Learning, 1–4, 109, 123, 145, 149–150,
169. *See also* Design, sociotechnical;
Error; Factories, sociotechnical;
Psychology; Workers
allocation of function and, 97
approach, technical basis for, 55–57
control and, 52–58, 73
integrating work and planning, 95–96
learning to learn, 95–97
pay-for-learning system, 2, 115, 127
preconception and habit, 88–90
sensory data, 96–97
skill, definition of, 91–93
synthetic vs. analytic reasoning,
90–91
and work, integration of, 113–123,
163
Lewin, Kurt, 118. *See also* Design,
sociotechnical

Linkage, mechanical. *See* Continuity
Little Chem, 122, 136. *See also* Design,
sociotechnical; Factories,
sociotechnical
Lorant, John H., 20, 46
Lotka, Alfred, 39

Machine design. *See* Design, machine;
Discontinuity in machine design
Machine systems. *See* Technology
Machine tooling, 8–9, 14, 19, 21–24
computer control of, 55
as cybernetic industry, 47–51
flexible manufacturing systems in,
48–50
machine-shop work culture, 154–155
metal-forming technologies, 153–154
systems impact on, 50–51
Management. *See also* Design,
sociotechnical; Error; Factories,
sociotechnical; Failure; Profits; Work;
Workers
careers, 165, 167, 168
game, the, 155–156
scientific, 10–11, 13–14, 61–72, 118,
157, 158 (*see also* Taylor, Frederick
Winslow)
in sociotechnical settings, 136–141,
145
teams, as failure mode, 165–166
Market. *See also* Citibank; Fiat; Olivetti
Company; Work, developmental
environment, new, 106–109
mass, 7, 8
Marx, Karl, 13, 18, 17ln
Massachusetts Institute of Technology,
47, 48
Maudslay, Henry, 17, 22
Maxwell, James, 33
Mayo, Elton, 65, 118
Mayr, Otto, 31
Mechanical technology. *See*
Technology, mechanical
Mechanization, 1, 5–15, 41, 42, 46. *See
also* Assembly line; Constraint;
Continuity; Labor, division of;
Standardization; Technology
Mechanization Takes Command, 7. *See also*
Giedion, Sigfried
Mistakes, worker. *See* Error
Modern Times, 12

Nelson, Daniel, 63
New Chem, 122. *See also* Design,
sociotechnical; Factories,
sociotechnical